Macromolecular Symposia

Symposium Editor: M.-A. De Paoli

Editor: I. Meisel
Associate Editors: K. Grieve, C.S. Kniep, S. Spiegel

189

pp. 1–141

November 2002

Macromolecular Symposia publishes lectures given at international symposia and is issued irregularly, with normally 14 volumes published per year. For each symposium volume, an Editor is appointed. The articles are peer-reviewed. The journal is produced by photo-offset lithography directly from the authors' typescripts.
Further information for authors can be obtained from:
Editorial office "Macromolecular Symposia"
Wiley-VCH
Boschstrasse 12, 69469 Weinheim,
Germany
Tel. +49 (0) 62 01/6 06-2 38 or -5 81; Fax +49 (0) 62 01/6 06-3 09 or 5 10;
E-mail: macro-symp@wiley-vch.de
http://www.ms-journal.de
Suggestions or proposals for conferences or symposia to be covered in this series should also be sent to the Editorial office at the address above.

Macromolecular Symposia:
Annual subscription rates 2003 (print only or online only)*
Germany, Austria € 1318; Switzerland SFr 2168; other Europe € 1318; outside Europe US $ 1568.
Macromolecular Package, including Macromolecular Chemistry & Physics (18 issues), Macromolecular Bioscience (12 issues), Macromolecular Rapid Communications (18 issues), Macromolecular Theory & Simulations (9 issues) is also available. Details on request.
* For a 5 % premium in addition to **Print Only** or **Online Only**, Institutions can also choose both print and online access.
Packages including Macromolecular Symposia and Macromolecular Materials & Engineering are also available. Details on request.
Single issues and back copies are available. Please, inquire for prices.

Orders may be placed through your bookseller or directly at the publishers:
WILEY-VCH Verlag GmbH & Co. KGaA, P. O. Box 10 11 61, 69451 Weinheim, Germany,
Tel. +49 (0) 62 01/6 06-4 00, Telefax +49 (0) 62 01/6 06-1 84, E-mail: service@wiley-vch.de

Macromolecular Symposia (ISSN 1022-1360) is published with 14 volumes per year by WILEY-VCH Verlag GmbH & Co. KGaA, P. O. Box 10 11 61, 69451 Weinheim, Germany. Air freight and mailing in the USA by Publications Expediting Inc., 200 Meacham Ave., Elmont, NY 11003. Periodicals postage pending at Jamaica, NY 11431. POSTMASTER: send address changes to Macromolecular Symposia, Publications Expediting Inc., 200 Meacham Ave., Elmont, NY 11003.

© WILEY-VCH Verlag GmbH & Co. KGaA, Weinheim, Germany, 2002
Printing: Strauss Offsetdruck, Mörlenbach. Binding: J. Schäffer, Grünstadt

**International Union
of Pure and Applied Chemistry**

Macromolecular Division

Polymer Science Insights

Invited lectures presented at the
6. Congresso Brasileiro de Polímeros
(6th Brazilian Polymer Conference)
held in
Gramado, Brazil
November 11–15, 2001

Symposium Editor
Prof. Marco-Aurelio De Paoli
Instituto de Quimica/UNICAMP
C. Postal 6154
13.083-970 Campinas, SP
Brasil
Fax: +55-19-3788-3023
E-mail: mdepaoli@iqm.unicamp.br

Congress Chair
Prof. Raquel S. Mauler, UFRS, Porto Alegre, Brazil

Scientific Committee
Ailton de Souza Gomes (IMA/UFRJ), Antonio Aprígio da S. Curvelo (IQSC/USP)
Inez V. P. Yoshida (IQ/UNICAMP), João Sinésio de C. Campos (FEQ/UNICAMP)
José Roberto M. D'Almeida (PUC/RJ)
Judith Pessoa de Andrade Feitosa (DQOI/UFC)
Laura Hecker de Carvalho (DEMa/UFPB), Marly M. Jacobi (IQ/UFRGS)
Rodrigo Oréfice (DEMM/UFMG), Sebastião V. Canevarolo Jr. (DEMa/UFSCar)
Susana A. Libermann (OPP Química), Wang Shu Hui (EP/USP/SP)

Contents of Macromol. Symp. 189

6th Brazilian Polymer Conference
Gramado (Brazil), 2001

Preface
M.-A. De Paoli

* The asterisk indicates the name of the author to whom inquiries should
  be addressed

Author Index

Preface

The "Congresso Brasileiro de Polímeros" (Brazilian Polymer Congress) is a biennial event which takes place at different locations in Brazil, near centers where groups active in polymer research exist. When it started, in 1991, the number of participants was small and most of them came from the universities and industries of the states of São Paulo and Rio de Janeiro. With time, the Brazilian community working on polymer research has expanded and other groups appeared, even in the most distant regions and states, from Rio Grande do Sul in the extreme south to Cear in the north of the country. This not only increased the number of participants but also the quality of the research presented at the meetings. Nowadays this Congress represents the most important meeting on polymer science in South America, involving participants from this continent as well as from other laboratories throughout the world. The support provided by IUPAC to the meeting held at Gramado, in November 2001, is an indication of the quality and acceptance of the meeting within the international community. This support includes the publication of this special issue of Macromolecular Symposia containing review articles written by the researchers who gave the plenary lectures, which covered all areas of interest to the participants.

M.-A. De Paoli

Campinas, Brazil

Hybrid Organic-Inorganic Polymer Networks Based on the Copolymerization of Methacryloxypropyl-Silsesquioxanes and Styrene

Patricia Eisenberg[1], Juan C. Lucas[1], Roberto J. J. Williams[2,] *

[1]Technological Research Center for the Plastics Industry (CITIP, INTI) and INDEMAT (University of San Martín), C.C.157, 1650 San Martín, Argentina
[2]Institute of Materials Science and Technology, University of Mar del Plata and National Research Council (CONICET), J. B. Justo 4302, 7600 Mar del Plata, Argentina

Summary: Silsesquioxanes synthesized by the condensation of (3-methacryloxypropyl)trimethoxysilane with concentrated formic acid, were copolymerized with styrene using benzoyl peroxide as initiator. When the ratio of styrene (S) to methacrylic (M) unsaturations was S/M > 1, the reaction attained an almost complete conversion as shown by the total reaction heat determined by differential scanning calorimetry. Several properties of the resulting hybrid organic-inorganic polymer networks were determined as a function of the silsesquioxane amount used in the formulation. Increasing the silsesquioxane fraction led to an increase of the glass transition temperature up to a point where devitrification was not observed prior to decomposition, an increase of the elastic modulus in the rubbery state, a decrease of the intensity of tan δ, and an increase of both Barcol hardness and thermal stability.

Introduction

Silsesquioxanes (SSO) are the products obtained by the polycondensation of monomers such as $RSiX_3$ (X = alkoxy, OH, Cl, etc., R = organic group), under appropriate acid or alkaline conditions. The resulting SSO consists of a distribution of species with a generic structure: $[RSiO_{1.5-x}(OH)_{2x}]_n$, where n may vary in a broad range. Multimodal distributions of molar masses are frequently observed, with relative maxima at n close to 8, 16, and higher values.[1,2] Depending on the selected reaction conditions, unhydrolyzed X groups (e.g., methoxy groups) may replace hydroxyls in the generic

2

structure.[3] Figure 1 shows typical structures of silsesquioxanes.

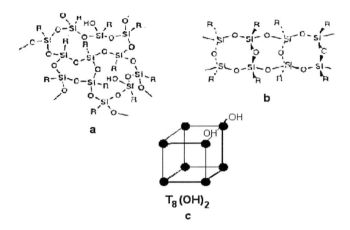

Figure 1. Typical structures of silsesquioxanes; a) random structure, b) ladder structure, c) incompletely condensed polyhedron (black circles represent Si atoms).

When the organic group R bears a reactive functionality (e.g., epoxy, vinyl, etc.), hybrid organic-inorganic polymer networks may be obtained by the addition of appropriate initiators, comonomers and catalysts.[4-9] Materials with a broad range of thermal and mechanical properties may be obtained by varying the relative proportions of the organic and inorganic parts.

The aim of this paper is to analyze thermal and mechanical properties of organic-inorganic hybrid materials obtained by the copolymerization of a methacryloxypropyl-functionalized SSO with variable amounts of styrene.

Experimental Part

Synthesis of the Silsesquioxane (SSO)

The SSO was obtained by polycondensation of (3-methacryloxypropyl)trimethoxysilane (Z-6030 Dow Corning), in bulk, with concentrated formic acid (Merck, 98 wt%). The reaction was carried out at 50 °C during about 14 days, leading to a homogeneous viscous liquid that could be completely dissolved in different solvents like tetrahydrofuran (THF), chloroform and styrene.

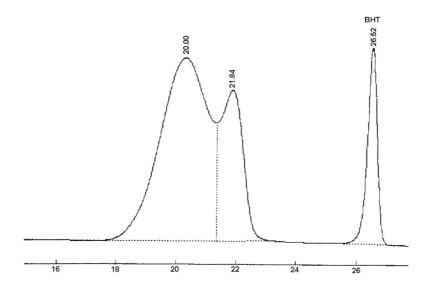

Figure 2. Size exclusion chromatogram of the silsesquioxane (butylated hydroxytoluene, BHT, was used as internal standard).

Size exclusion chromatograms (Shimadzu GPC 80, with sim-pack Shimadzu columns 801, 802 and 803, refractive index detector and THF as carrier at 1 ml/min), showed the presence of a bimodal molar-mass distribution (Figure 2). From a calibration with polystyrene standards (Shodex, range of molar masses from 10^3 to 6×10^6), the peak at 21.84 min was assigned to species with $n = 6\text{-}11$, while predominant species in the peak at 20.0 min corresponded to n values comprised between 17-20. Mass spectrometry confirmed the assignment and showed that these oligomers exhibited a high fraction of intramolecular cycles.[10] The average structural formula of the SSO could be represented by $(RSiO_{1.42}(OH)_{0.16})_n$, where $R \cdot CH_2=C(CH_3)COO(CH_2)_3$. This gives a mass of 180.4 g/mol of C=C double bonds.

Copolymerization with Styrene

Benzoyl peroxide (BPO) was dissolved in styrene, in an amount equal to 1.9 wt % of total organic mass (including the mass of the R group of the SSO, added in a subsequent

4

step). The SSO was then dissolved, giving a solution characterized by a particular value of the wt % styrene. The free-radical polymerization was performed under two different conditions: a) at 10 °C/min up to 180 °C, under N_2, in a differential scanning calorimeter (Mettler TA 3000 DSC-20); b) casting the solution into a mold consisting of two glass plates coated with siliconized paper, spaced by rubber cords 2 mm in diameter, and heating in an oven at 80 °C during 17 h, with a postcure step at 160 °C during 3h. The resulting hybrid organic-inorganic materials were transparent. Hybrids containing less than about 30-40 wt % styrene were very brittle, making it very difficult to demold the part without breaking it. Cured samples containing different amounts of styrene, were kept in THF during 4 days at room temperature, plus another 2 hours in an ultrasonic bath. No loss of mass was observed after this period, meaning that the copolymerization of the SSO and styrene had been effective.

Characterization Techniques

The total reaction heat was obtained using differential scanning calorimetry (Mettler TA 3000 DSC-20). Dynamic scans were performed at 10 °C/min up to 180 °C, under N_2.

Specimens, about 10x10x2 mm in size, were machined from the molded plaques and subjected to single cantilever bending in a DMTA IV Rheometrics, under the following conditions: scanning rate = 3 °C/min, frequency = 1 Hz, maximum strain = 0.1%.

The Barcol hardness of plaques (ASTM D 2583), was measured with a Portable Hardness Tester (Barber-Colman, Model 934-1).[11]

Thermal gravimetric analysis (TGA, Mettler TG 50) was carried out under nitrogen flow at 10 °C/min up to 800 °C. At this temperature, nitrogen was replaced by oxygen and heating was continued up to 900 °C, attaining a constant residual mass. This enabled the ceramic yield to SiO_2 to be determined.[5]

Results and Discussion

Total Reaction Heat

Figure 3 shows experimental values (circles) of the total reaction heat obtained by differential scanning calorimetry, as a function of the wt% of styrene in the initial formulation. Squares represent theoretical values expected for complete conversion of vinyl groups of both styrene and the silsesquioxane.

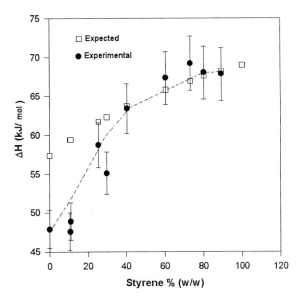

Figure 3. Experimental values (circles) of the total reaction heat obtained by differential scanning calorimetry, as a function of the wt % styrene in the initial formulation. Squares represent theoretical values expected for complete conversion of vinyl groups of both styrene and the silsesquioxane.

A value of 69 kJ/mol was taken for styrene (S) and 57.4 kJ/mol for the vinyl groups of the SSO, considered equal to the value reported for the polymerization of n-propyl methacrylate.[12]

For a wt % S higher than about 30-40, experimental values of reaction heats were close to expected values within experimental error. For a wt % S = 36.5 both monomers were present in equimolar amounts. This means that the necessary condition to get a complete conversion, is to add one mol of styrene per mol of methacryloxy groups of the SSO. For a smaller amounts of styrene, the reaction could not attain complete conversion, probably due to steric reasons. For the pure SSO, the maximum conversion of vinyl groups was 83% of the expected value.

Trends found for this particular system agree with experimental results reported for other hybrid polymers. For example, quantitative analysis of the photochemical polymerization reaction of an hybrid polymer obtained by the hydrolytic condensation of a 1:1 mixture of (3-methacryloxypropyl)trimethoxysilane and Si(OEt)$_4$, revealed that about 20% of the initial double bonds remain unreacted. However, addition of 2 moles of methyl methacrylate per mol of methacryloxy groups of the hybrid polymer, led to a complete conversion of double bonds upon UV irradiation.[13]

Dynamic Mechanical Characterization

Figure 4 shows the storage modulus of cured hybrids formulated with different SSO amounts, as a function of temperature. The curve for an injection-grade polystyrene is also included for comparison purposes. Increasing the amount of SSO up to 30 wt % led to an increase of the glass transition temperature and the rubbery modulus, due to the increase in crosslink density. For higher amounts of SSO, hybrids were very brittle. This led to the generation of microcracks that eventually produced the material's fracture in the course of the test. One of these runs is shown in Figure 4, for a hybrid containing 90 wt % SSO. Although a small decrease of the glassy modulus with temperature, was observed, devitrification did not take place in the temperature range preceding thermal degradation (small step-decreases of the storage modulus were assigned to the presence of microcracks).

An explanation of the significant dependence of the dynamic-mechanical behavior of the synthesized hybrids with their composition, may be postulated on the basis of recent results reported by Matejka et al.[14] By analyzing the structure of several silsesquioxanes using small-angle X-ray scattering, they found that because of the incompatibility of the polyhedral SSO framework and organic chains, microphase separation took place and spontaneous self-organization occurred. A micellar arrangement of compact SSO domains with a correlation distance corresponding to the size of the substituent, was observed. In our case, it may be postulated that a self-organization was present before polymerization and it was frozen after formation of the polymer network. For hybrids with low SSO mass fractions, the organic microphase may be considered as the continuous one, and the materials exhibited a characteristic T_g that increased with crosslink density.

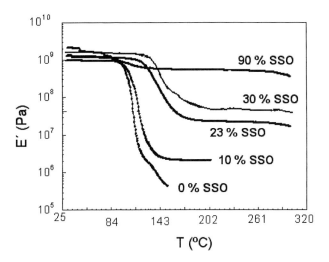

Figure 4. Storage modulus of cured hybrids formulated with different wt % SSO, as a function of temperature.

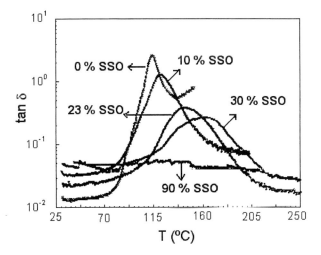

Figure 5. Evolution of tan δ for the same hybrids of Figure 4.

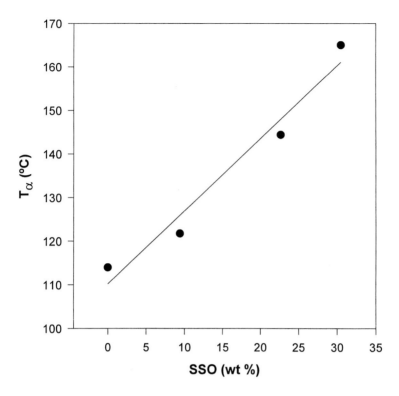

Figure 6. T_α relaxation, defined at the maximum of tan δ, as a function of the wt % of SSO in the hybrid material.

When the SSO mass fraction is high, the rigid inorganic microphase percolates through the structure, leading to the appearance of a glassy behavior up to the degradation temperature. Hybrids with a high SSO content might be used as coatings with a thickness small enough to avoid manifestation of the brittle behavior.

Figure 5 shows the evolution of tan δ for the same hybrids of Figure 4 (including the neat polystyrene for comparison purposes). Increasing the amount of SSO led to a significant decrease in the intensity of tan δ, associated with an increase in the rigidity of the network structure. For the hybrid with 90 wt % SSO, no neat network relaxation could be observed in the whole temperature range.

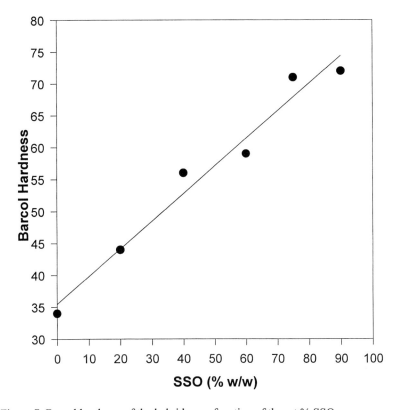

Figure 7. Barcol hardness of the hybrids as a function of the wt % SSO.

The T_α relaxation, defined at the maximum of tan δ, is plotted in Figure 6 as a function of the wt % of SSO in the hybrid material. For a 30 wt % SSO, T_α (associated with the glass transition temperature), increased by about 50 °C with respect to pure polystyrene.

Hardness

The Barcol hardness of the hybrids increased linearly with the wt % SSO, as shown in Figure 7. This resulted from the relative contribution of the individual hardnesses of both microphases, present in the glassy state at room temperature. For comparison purposes, the hardness of a typical crosslinked unsaturated polyester was measured with the same test and found to be close to that of the hybrid with 40 wt % SSO.

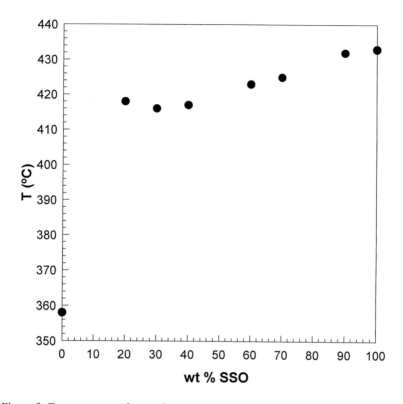

Figure 8. Temperature at the maximum rate of thermal degradation as a function of the wt % SSO in the hybrid material.

Thermal Stability

It has been reported that thermoplastics modified with polyhedral silsesquioxanes exhibit an increase in their decomposition temperature and are very resistant to combustion. This property has been associated with the preceramic nature of the silsesquioxane component, which forms a glassy layer of SiO_xC_y during pyrolysis, retarding the diffusion of gases through the surface char.[15] For example, the decomposition temperature of poly(4-methylstyrene), defined as the temperature at which 10 % weight loss was observed when heating a powdered sample under nitrogen at 10 °C/min, was 388 °C.[16] The copolymerization of 4-methylstyrene with an octahedral silsesquioxane bearing one polymerizable styryl group, led to hybrid organic-

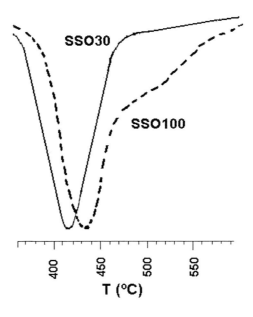

Figure 9. Rate of mass variation by thermal degradation, dm/dT, as a function of temperature for a hybrid with 30 wt % SSO and for the neat SSO.

inorganic thermoplastics. Their decomposition temperature increased with the SSO amount, attaining values as high as 445 °C for the pure poly(silsesquioxane) (the particular value depended on its molar mass).[16]

In our case, a polymer network was generated by copolymerization of the multifunctional SSO species with styrene. Figure 8 shows the temperature corresponding to the maximum rate of thermal degradation under nitrogen, as a function of the wt % SSO in the hybrid material. The degradation temperature increased from 358 °C for pure PS, to about 417 °C for hybrids with 20-40 wt % SSO, and to about 435 °C for the pure SSO. A significant increase in the thermal stability of resulted from the incorporation of only 20 wt % SSO. Possibly, network formation produced an extra increase of thermal stability when compared with SSO-modified thermoplastics.

Examples of the derivatives of the mass loss, dm/dT, as a function of temperature are shown in Figure 9 for a hybrid with 30 wt % SSO and for the neat SSO. The shoulder observed in the high-temperature range for the neat SSO, may be ascribed to the formation of a SiO_xC_y glassy layer that retards further decomposition.[15]

12

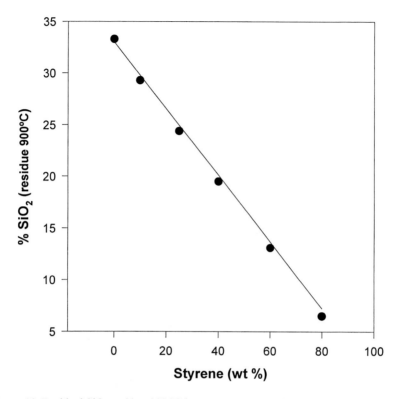

Figure 10. Residual SiO_2 wt % at 900 °C in an oxygen atmosphere, compared with the expected value (full line), for hybrids containing different styrene amounts.

The ceramic yield to SiO_2 was obtained by heating the hybrids to 900 °C in an oxygen atmosphere. Figure 9 shows the residual SiO_2 wt % obtained for the different hybrids, compared with the expected values (full line). A very good agreement was observed.

Conclusions

Hybrid organic-inorganic materials based on a silsesquioxane functionalized with methacryloxy groups, copolymerized with styrene, exhibited a broad range of thermal and mechanical properties, depending on the weight fraction of SSO in the formulation. Increasing the SSO amount led to an increase of the glass transition temperature (T_α

relaxation), the elastic modulus in the rubbery state, hardness and thermal stability. Dynamic-mechanical properties were explained by assuming a self-organization of the structure in organic and inorganic domains, as found for several silsesquioxanes.[14] Percolation of the inorganic domains occurs at large SSO fractions and provokes the disappearance of the glass transition temperature. In these conditions, the hybrid behaves as a brittle glass up to the degradation temperature. The thermal stability of the hybrid materials increased significantly with the SSO content. This was ascribed to the formation of a SiO_xC_y glassy layer that retarded the degradation process.

References

[1] K. Piana, U. Schubert, *Chem. Mater.* **1994**, *6*, 1504.

[2] P. Eisenberg, R. Erra-Balsells, Y. Ishikawa, J. C. Lucas, A. N. Mauri, H. Nonami, C. C. Riccardi, R. J. J. Williams, *Macromolecules* **2000**, *33*, 1940.

[3] R. J. J. Williams, R. Erra-Balsells, Y. Ishikawa, H. Nonami, A. N. Mauri, C. C. Riccardi, *Macromol. Chem. Phys.* **2001**, *202*, 2425.

[4] U. Schubert, N. Hüsing, A. Lorenz, *Chem. Mater.* **1995**, *7*, 2010.

[5] A. Sellinger, R. M. Laine, *Macromolecules* **1996**, *29*, 2327.

[6] P. Eisenberg, J. C. Lucas, R. J. J. Williams, *Polimery(Warsaw)* **1999**, *44*, 735.

[7] K. H. Haas, *Adv. Eng. Mater.* **2000**, *2*, 571.

[8] A. N. Mauri, C. C. Riccardi, R. J. J. Williams, *Macromol. Symp.* **2000**, *151*, 331.

[9] A. N. Mauri, C. C. Riccardi, R. J. J. Williams, *Polym. Bull.* **2001**, *45*, 523.

[10] P. Eisenberg, R. Erra-Balsells, Y. Ishikawa, J. C. Lucas, H. Nonami, R. J. J. Williams, *Macromolecules* **2002**, *35*, 1160.

[11] V. Shah, *"Handbook of Plastics Testing Technology"*, Wiley, New York 1984.

[12] R. M. Joshi, *Makromol. Chem.* **1963**, *66*, 114.

[13] L. Delattre, C. Dupuy, F. Babonneau, *J. Sol-Gel Sci. Technol.* **1994**, *2*, 185.

[14] L. Matejka, O. Dukh, D. Hlavatá, B. Meissner, J. Brus, *Macromolecules* **2001**, *34*, 6904.

[15] J. D. Lichtenhan, *Comments Inorg. Chem.* **1995**, *17*, 115.

[16] T. S. Haddad, J. D. Lichtenhan, *Macromolecules* **1996**, *29*, 7302.

Macromol. Symp. **189**, 15–26 (2002)

Polymer Electrostatics: Detection and Speciation of Trapped Electric Charges by Electric Probe and Analytical Electron Microscopy

Fernando Galembeck, Carlos Alberto Paula Leite,
Maria do Carmo V. M. da Silva, Amauri José Keslarek,
Carlos Alberto Rodrigues Costa, Érico Teixeira-Neto,
Márcia Maria Rippel, Melissa Braga

Instituto de Química, Universidade Estadual de Campinas, PO Box 6154, Campinas SP, Brazil

Summary: This work reviews new probe and electron microscopy approaches for the detection of charged domains in insulating polymers, as well as for the identification of the charge-bearing species: scanning electric potential microscopy (SEPM), electric force microscopy (EFM) and energy-loss spectroscopy imaging in the transmission electron microscope (ESI-TEM). The SEPM and EFM micrographs show patterned domains bearing excess electric charges and extending for tens of nanometers, in polymer latex particles and films. The charged species are identified by ESI-TEM as emulsion polymerization initiator and surfactant residues, as well as the associated counter-ions. Charged domains are also observed in common thermoplastic polymers, producing unexpectedly large electric potential gradients.

Introduction

Polymer dielectrics are usually described as electro-neutral solids, even though there is plenty of evidence showing that they contain domains with excess electric charges. Several phenomena related to the existence of fixed charges or polarization in polymers are described in the literature: their ability to acquire charges by friction, the formation of electrets, space and residual charges, double-layers at interfaces and interfacial polarization, which were shortly reviewed in a recent paper.[1] However, there is a conceptual constraint in dealing with these problems, that is the current idea of charge neutrality as a normal state for polymer dielectrics at every relevant length scale, from the macromolecules up to the macroscopic plastic solids (films, coatings, tubes) and devices. Of course, the prevalence of electroneutrality in any physical medium suggests that the ionic species carrying opposite charges should occupy neighboring sites and they should also move simultaneously.[2] A

          CCC 1022-1360/00/$ 17.50+.50/0

review article on the non-linear optical polymer electrets acknowledges that a corona discharge produces and carries chemically active species, which tend to attack and modify the surface and subsurface layers of organic materials. However, the spatial distribution and chemical identity of the resulting charge species responsible for polymer polarization was not described.[3] Other authors acknowledge the existence of ionic carriers in insulating polymers, but these have not yet been identified or located.[4]

These difficulties have not prevented the use of thermally stimulated discharge currents in polymers, in the study of polymer relaxations,[5,6,7] as well as the recognition of the formation of double-layers of opposite electrical charges by two dielectric contacting phases.[8,9]

Indeed, the formation of an electrical double layer at an interface is expected, considering the differences of dielectric constant between the two adjoining phases. Since most crystalline polymers are indeed multiphase systems, they should intrinsically display internal charge separation, due to the Maxwell and the Costa Ribeiro[10,11] effects. This idea is further supported by evidences[12] showing that the crystalline-amorphous interface plays an important role not only as a charge-trapping site but also in carrying an important amount of the total polarization in PVDF. The differences in dielectric constants of *e.g.* crystalline and amorphous domains in polyethylene (and other partially crystalline thermoplastics) are probably small, but their electronic polarization component is sufficient to impair the transparency of the partially crystalline polymer.

Beyond the differences of dielectric constant between amorphous and crystalline polymer domains, there are other factors that influence the formation of differentiated domains within a polymer: oxidized polymer chains can be segregated, as well as the immiscible catalyst residues and contaminants introduced during the fabrication of polymer and artifacts. Polymers prepared by emulsion polymerization or processed in polar media should always be contaminated with trapped ions, and these are not necessarily arranged in electro-neutral ion pairs or clusters.

However, there is also a prevailing but unproven idea, following which the charge-bearing species in neutral solids are found in ion pairs or clusters and thus they do not generate any effects at larger distances. This idea is largely due to a major experimental limitation, namely the difficulty to determine the existence of ionic species within insulating solids.

This difficulty is being overcome with the development of new techniques for electric potential or force mapping across solid surfaces, using a variety of scanning probe microscopies, appropriate for sensing charges,[13] dielectric constants, film thickness of insulating layers, photo-voltage and electric potentials.[14, 15]

Excess charges are necessarily associated with ionic species or free electrons, but charge speciation in dielectrics is also a challenging, largely unsolved problem of condensed matter physics and chemistry, even in the seemingly simpler case of liquid dielectrics.[16] This problem is addressed using an analytical electron microscopy; the method of choice is electron energy-loss spectroscopy associated to transmission electron microscopy (ESI-TEM), due to the high sensitivity coupled to a sub-nanometer spatial resolution. The present work reviews the application of microscopy to the detection of domains with excess electric charges, in insulating polymers.

Probe microscopy for electric domain mapping: SEPM and EFM

Scanning electric potential microscopy (SEPM) images are acquired using the standard non-contact AFM set-up (in this case, a Topometrix Discoverer instrument) but with the following modifications: the probe tip is coated with Pt and it is fed with an AC signal, 10 kHz below the frequency of the normal AFM oscillator, which matches the natural frequency of mechanical oscillation of the cantilever-tip system (40-70 kHz). During a measurement, the mechanical oscillation of the tip is tracked by the four-quadrant photodetector and analyzed by two feedback loops. The first one is used in the conventional way to control the distance between tip and sample surface, while scanning the sample at constant oscillation amplitude. The second loop is used to minimize the electric field between tip and sample: a second lock-in amplifier measures the tip vibration at the AC frequency oscillation while scanning, and adds a DC bias to the tip, to recover the undisturbed AC oscillation. This technique differs from that used by Terris,[10] who measured the phase displacement of the AC voltage while in the Topometrix set up the phase displacement is cancelled by DC biasing. The image is built up by using the DC voltage fed to the tip, at every pixel, thus detecting electric potential gradients throughout the scanned area. This technique derives from the Kelvin bridge and it is reminiscent of the oscillating electrode method for monolayer study: both use an oscillating electrode separated from the sample by an air gap. The main difference between them is the detection technique used, since SEPM uses a phase detection of the voltage applied.

The electric force microscopy (EFM)[17] technique uses a different approach: the surface is scanned at two different constant heights, typically 10 and 70 nm. In the first case, the interaction between tip and sample is determined by the short-range van der Waals forces, but at 70 nm the interaction is dominated by the electrostatic interactions. This technique is easily performed, but the results are not interpreted as unequivocally as the SEPM data.

A typical SEPM result is presented together with an AFM image of the same sample area in Figure 1, for a polystyrene latex. This was prepared by emulsion polymerization initiated with K_2S_2O in the presence of a mixture of polyoxyethylene-23-lauryl ether and sodium dodecylsulfate surfactants.[18]

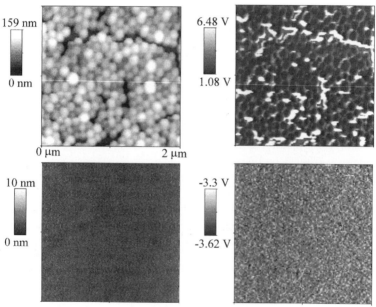

Figure 1. SEPM (right) and AFM (left) images of PS latex films on mica. On top are the figures of the as-prepared dry particles, and on the bottom is the film resulting from exposure of a dry film to a chloroform-saturated atmosphere, for 12 hours.

Another pair of AFM and SEPM images in the same figure show a film of the same latex, but after exposure to an atmosphere saturated with chloroform.

Due to the polystyrene high T_g, the particles do not coalesce in the dry film; their cores are negative relative to the outer shells and to the inter-particle spaces, where the (positive) counter-ions are expected to accumulate. The observed potential differences exceed 5 volts, corresponding to large potential gradients. Following exposure to chloroform, the film appears very flat (maximum height difference is 10 nanometers in the area shown) and the observed potential differences are much smaller, less than 0.5 V but still larger than the zeta potential for this latex, which is -29 mV. [19]

AFM and SEPM images for a film of a low T_g (15°C) poly(styrene-butyl acrylate-acrylic acid) latex prepared using redox initiation and sodium phenylether phosphate surfactant [20] have some unexpected features, as shown in Figure 2. Due to its low T_g value, films cast from this latex are transparent, suggesting that particles are very well coalesced. However, particle borders are easily seen, both in the AFM and SEPM images.

This charge pattern is similar to that of the natural rubber latex (NRL) film in Figure 3: NRL also forms transparent films at room temperature, but particle individuality is easily observed at the film surface, even though the T_g of cis(polyisoprene) is very low.

Figure 2. AFM and SEPM images of a transparent (transmittance>99%) copolymer (styrene-butyl acrylate-acrylic acid) latex film.

Consequently, there is a limit to inter-particle diffusion, and this is likely due to the particle shells with a different chemical composition from the particle bulk as evidenced by the

20

large positive electric potential, observed in the SEPM picture in Figure 3.

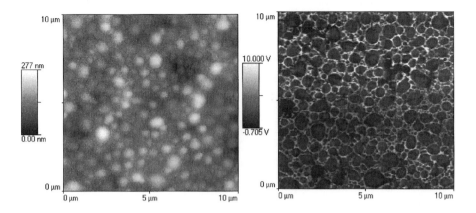

Figure 3. Non-contact AFM and SEPM maps of a natural rubber latex film, cast from freshly collected, centrifuged latex.

All these images have to be examined considering that the electrostatic interactions are long-range. This means, charges buried beneath the sample surface for tens and even a few hundred nanometers are detectable, but their contribution to the measured potentials decreases with the square of their distance to the scanning probe. In the case of the latexes, the observed electric potential gradients can be assigned to a spatial separation between anions (arising *e.g.* from the persulfate initiators or from anionic surfactants) and the corresponding counter-cations.

Analytical electron microscopy: ESI-TEM

The elemental distribution within fine latex particles and thin films can be observed using the electron spectroscopy imaging technique, in a transmission electron microscope (ESI-TEM). A Carl Zeiss CEM 902 transmission electron microscope, equipped with a Castaing-Henry-Ottensmeyer energy filter spectrometer within the column was used. The spectrometer exploits inelastic scattered electrons to form element-specific images. When the electron beam passes through the sample, interaction with electrons of different elements results in characteristic energy losses. A prism-mirror system deflects electrons with different energies to different angles so that only electrons with a well-defined energy are selected. If elastic electrons are the only chosen ($\Delta E = 0$ eV) a transmission image with reduced chromatic aberration is obtained. When monochromatic inelastic scattered electrons are selected, electron spectroscopy images (ESI) are formed, in which contrast is dependent on the local concentration fluctuations of a particular chosen element. Clear areas correspond to element-rich domains.

Examination of individual latex particles is possible, provided they are sufficiently small, or they are spontaneously flattened over the supporting carbon film. For larger particles, ultra-microtome cuts are used, [21] to observe the elemental distribution in the particle bulk.

This technique produced some unexpected results. For instance, in many types of latex the sulfur elemental map shows this element distributed across the particle bulk but depleted in the surface. Since sulfur is found as sulfate or sulfonate from initiator residues or surfactant, these particles have negative charges throughout their interior. On the other hand, the sodium and potassium counter-ions are often observed in a narrow shell surrounding the particle, but in other cases they appear spread throughout the whole particle volume. This means that most dry latex particles have a negative core and a positive shell. Moreover, since the charge distributions are not highly symmetric, the particles are electric multipoles, and the domains with excess positive or negative charges extend for many tens and even hundreds of nanometers.

More recently, rather complex elemental distribution patterns have been observed, as shown in Figure 4. These elemental maps are from a submonolayer of the same particles presented in Figure 2. The individual particles are strongly deformed and flattened over the supporting carbon film, allowing the observation of a strong accumulation of P and thus of the phosphate groups bearing negative charges, at the particle borders. These P-rich domains are also electron-dense, since they are observed as very dark spots in the bright-field image. On the other hand, they appear as darker areas in the carbon elemental map, showing that they have a low carbon content relative to the particle bulk. O is distributed throughout the particles, since it is largely associated with the acrylic monomer.

Spectroscopy imaging in the transmission microscope is a *transmission* technique; consequently the images show the presence of an element throughout the whole sample thickness, provided the sample is sufficiently thin to allow for the escape of a few inelastic

22

scattered electrons. In another frequent pattern, sulfur (and thus the negative sulfate groups)

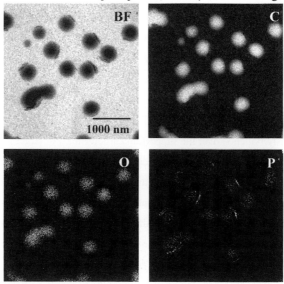

Figure 4. Bright-field (BF) picture and elemental (carbon, C; oxygen, O; phosphorus, P) maps of a poly(styrene-co-butyl acrylate-co-acrylic acid) latex ($T_g=15^\circ$C).

is unevenly distributed throughout the whole particle, as shown in the elemental maps for a polystyrene latex in Figure 5.

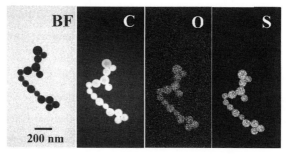

Figure 5. Bright-field picture and elemental maps of a poly(styrene) latex. This is the same latex presented in Fig. 1.

Charge mapping in thermoplastics

A previous work[1] presented electrical distribution maps for a variety of common thermoplastics, and this was later observed also in the case of plasticized cellulose acetate membranes prepared following a phase-inversion procedure [22] (Figure 6).

Understanding the large electric potential gradients in the cellulose acetate, polyolefin and other SEPM pictures is not as easy as in the case of the latexes, because they cannot be associated to local excess of ionic charge. The acetylated cellulose can bear a few carboxylic groups arising from cellulose oxidation, but cellulose acetate membranes are well known for their low adsorption ability, which makes them highly suitable for the handling of proteins and other biological materials. This low adsorption ability is in turn evidence for the absence or a very low concentration of ionic groups, in the membrane surface. Consequently the large observed electric potential gradients can hardly be assigned solely to the oxidized cellulose residues.

Before raising other hypotheses, it should be recalled that the species responsible for charge separation have not yet been properly described, regardless if Maxwell or Costa Ribeiro effects is concerned.

A possible explanation for the charge separation in cellulose acetate membranes is based on the well-known ability of hydrophobic surfaces for differential adsorption of H^+ and OH^- ions, with an advantage to the later. Consequently, the more hydrophobic surface domains at the polymer surface during membrane casting could be loaded with hydroxyl ions, as opposed to the more hydrophilic domains, thus creating spots with negative potential. Other possible explanations are the electro-kinetic phenomena associated to mass transfer during film casting, *e.g.* streaming potentials and junction potentials. These are usually less than 100 mV when the polymer is highly swollen with water, but may reach much higher values upon drying, when the separated charges are trapped in liquid pools and finally they are rendered immobile.

Still another possibility for charge separation arises from tribochemistry [23]: chain rupture in the stressed quasi-dry membrane can be at least partly heterolytic, thus producing chain - bound ions. The demonstration of tribochemical free radical formation is rather easy by electron-spin resonance, but unfortunately the formation of ionic defects cannot be monitored by highly sensitive fast spectroscopic techniques, analogous to free radicals.

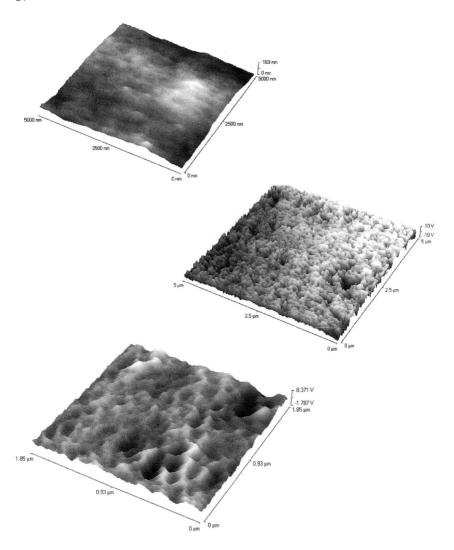

Figure 6. AFM (top) and SEPM (medium and bottom) images of a cast cellulose acetate membrane prepared by phase inversion and plasticized with glycerol, prior to drying and storage. Acquisition of the dry membrane images was not possible; so the membrane was wetted with 1.0×10^{-3} mol L^{-1} aqueous KCl, prior to the acquisition of these images.

It is important to verify the significance of local electric potentials experimentally determined close to film surfaces, by comparing them to the results of model calculations. This is being done in the authors' laboratory, by evaluating the local potentials generated by predefined distributions of electric charges. For instance, a 400-nm diameter disc with 176 equally spaced negative charges and surrounded by a 20-nm thick shell studded with 176 positive charges generates an electric potential map, measured 10 nm above the surface, well represented as a positive ring enclosing a negative disk. This is consistent with the images in Figures 1-3 and with the SEPM images for poly(styrene-co-hydroxyethylmetacrylate) films, published elsewhere.[1]

Conclusion

Polymer films and particles are decorated with patterns of charged domains, as evidenced by new electric probe microscopy techniques, as well as by energy-loss analytical electron microscopy. Polymer colloidal particles and thin films charged domains are associated with the local concentrations of ionic groups, but the speciation of charge-bearing groups in cellulose acetate and other common thermoplastics has not yet been achieved, and it will depend on the combination of electric probe microscopy with a spatially-resolved molecular spectroscopy, perhaps near-field Raman imaging.

The mosaic of positive and negative electric potentials detected just above the polymer surface proves the existence of an electric charge mosaic within the polymer. This charge distribution is not explicitly recognized in the current polymer literature, and some currently widespread assumptions and models are actually in disagreement with the experimental data presented in this paper.

For these reasons, future work will have three major aims:

i) evaluation of the effect of excess charges on polymer mechanical, optical, permeation, adhesion and chemical properties.

ii) identification of other species beyond simple ions, in the domains with different local charges, probably by using near-field Raman spectroscopy.

iii) use of this new information on charged sites in polymers, to develop new polymer materials.

Acknowledgements: AJK, ETN, MB and MMR are graduate fellows from Fapesp. FG acknowledges support from CNPq, Fapesp, Pronex/Finep/MCT, and the Serrana company. This is a contribution of Millenium Institute for Complex Materials (PADCT/MCT).

[1] A. Galembeck, C. A. R. Costa, M. C. V. M. Silva, E. F. Souza, F. Galembeck, *Polymer* **2001**, *42*, 4845.
[2] K. Wu, M. J. Iedema, J. P. Cowin, *Science* **1999**, *286*, 2482.
[3] S. Bauer-Gogonea, R. Gerhard-Multhaupt, *IEEE Trans Dielectr Electr Insul* **1996**, *3*, 677.
[4] J. A. Malecki, *Phys Rev B* **1999**, *59*, 9954.
[5] C. Lacabanne, P. Goyaud, R. F. Boyer, *J Polym Sci* **1980**, *18*, 277.
[6] K. Shrivastava, J. D. Ranade, A. P. Srivastava, *Thin Solid Films* **1980**, *67*, 201.
[7] M. Mudarra, J. Belana, J. C. Cañadas, J. A. Diego, *Polymer* **1999**, *40*, 2569.
[8] S. M. Skinner, R. L. Svage, J. E. Rutzler Jr, *J Appl Phys* **1953**, *24*, 438.
[9] S. M. Skinner, *J Appl Phys* **1955**, *26*, 509.
[10] J. C. Ribeiro, *An Acad Bras Cienc* **1950**, *22*, 325.
[11] P. Eyerer, *Adv Colloid Interface Sc* **1972**, *3*, 223.
[12] D. Rollik, S. Bauer, R. Gerhard-Multhaupt , *J. Appl Phys* **1999**, *85*, 3282.
[13] B. D. Terris, J. E. Stern, D. Rugar, H. J. Mamin, *J Vac Sci Technol A-Vac Surf Films* **1980**, *8*, 374.
[14] M. Nonnenmacher, M. P. O'Boyle, H. K. Wickramasinghe, *Appl Phys Lett* **1991**, *58*, 2921.
[15] F. Saurenbach, B. D. Terris, *Appl Phys Lett* **1990**, *56*, 1703.
[16] W. F. Schmidt, "*Liquid State Electronics of Insulating Liquids*", CRC Press, Boca Raton NY **1997**.
[17] F. Galembeck, C. A. R. Costa, A. Galembeck, M. C. V. M. Silva, *An Acad Bras Cienc* **2001**, *73*, 495.
[18] J. M. Moita-Neto, V. A.R. Monteiro, F. Galembeck, *Colloids Surf A* **1996**, *108*, 83.
[19] M. Braga, C. A. R. Costa, C. A. P. Leite, F. Galembeck, *J Phys Chem B* **2001**, *105*, 3005.
[20] A. J. Keslarek, C. A. R. Costa, F. Galembeck, *Langmuir* **2001**, *17,* 7886.
[21] A. H. Cardoso, C. A. P. Leite, F. Galembeck, *Langmuir* **1998**, *14*, 3187.
[22] S. P. Nunes, F. Galembeck, *J Polym Sci Part C - Polym Lett* **1983**, *21*, 49.
[23] G. Heinicke, "*Tribochemistry*", Hanser, Berlin **1984**.

Porous Crosslinked Spherical Resins for Diversified Applications: Packing Materials for Size Exclusion Chromatography

Fernanda M.B. Coutinho*[1,2], Márcia A.F.S. Neves[2] , Marcos L. Dias[2]

[1]Departamento de Processos Químicos, Instituto de Química, Universidade do Estado do Rio de Janeiro, Rio de Janeiro, RJ, Brazil, **fern@uerj.br*
[2]Instituto de Macromoléculas Professora Eloisa Mano, Universidade Federal do Rio de Janeiro (IMA/UFRJ), 21945-970, P.O. Box 68525 - Rio de Janeiro, RJ, Brazil, **fern@ima.ufrj.br*

Summary: Improvements in the synthesis of porous polymers for different applications have been carried out in our laboratory. Beads of poly(styrene-co-divinylbenzene) with morphology adequate to the application at hand were prepared. Packing materials for size exclusion chromatography (SEC) were prepared by single-step swelling and polymerization (SSWP) and by modified suspension polymerization (MSP). High values of exclusion limit (10^6 and 8.0×10^6) were attained for SEC columns packed with poly(styrene-co-divinylbenzene) synthesized using high proportions of polystyrene, as porogen agent, and divinylbenzene. The maximum values of exclusion limits were attained for SEC columns packed with beads prepared by SSWP method.

Introduction

The growing interest for porous polymer beads in the last three decades is due to their great number of diversified applications as packing materials for size exclusion chromatography (SEC), as supports for solid-phase organic synthesis, polymer-supported catalysts, polymer-immobilized extractants, adsorbents, ion exchange and chelating resins, etc. These polymers are produced by free radical aqueous suspension copolymerization of an adequate monovinyl monomer with a divinyl monomer or by chemical modification of a monovinyl-divinyl copolymer.[1-6]

A great number of ion-exchangers and polymer reagents are based on styrene-divinylbenzene copolymers (supports) which characteristics depend on the support morphology and on the morphological changes of its three-dimensional network after introducing the active sites. The specified application of these copolymers are closely related to their morphology and swelling properties.[7]

The morphology of these copolymers is governed by the occurrence of phase separation during the copolymerization reaction. That process is achieved by combining high proportion of crosslinking monomer and porogens in the polymerization mixture. Typical porogens employed in suspension polymerization can be classified into three types: solvating diluents (thermodynamically good solvents); non-solvating diluents (thermodynamically poor solvents); soluble polymers or mixtures thereof. At the end of the polymerization process the porogen agent is removed from the polymer network and the porous structure is attained. The type and amount of porogen and the proportion of crosslinking monomer relative to the vinyl monomer are the most important reaction variables in determining the type of morphology produced, that may vary from gel-type to macroporous with many degrees between them.[4] The gel-type resins do not have pores in dry state, they must be swollen in a good solvent in order to present porous properties.[8]

In the macroporous polymers, the macropores are often accessible to both solvents and non-solvents and in general the whole structure is not susceptible to the dramatic changes observed with gel-type resins when the nature of the surrounding medium is changed. However, the choice of using a gel-type or a macroporous polymer depends very much on the final application. In a swelling solvent (good solvent) gel-type beads can often be loaded or chemical modified to higher levels than macroporous ones. Nevertheless, the latter are generally much less sensitive to the choice of solvent to be used in the process at hand. Swollen gel-type resins are less affected to sudden shock but cannot be subjected to steady and high pressures. In opposition, the rigid macroporous beads are brittle and fracture under sudden stress, but can withstand considerable steady pressures.[8]

The objective of this paper is to describe the synthesis of porous beads with small particle size and low size distribution with varied morphologies adequate as packing material for SEC columns.

General procedures for the synthesis of porous polymer beads

Polymer beads are synthesized by free radical initiators through a suspension polymerization process. That process produces spherical particles with sizes dependent on the polymerization conditions. Styrene and acrylic monomers and divinylbenzene, as crosslinking agent, are among the most employed monomers because of the variety of reactions which can be carried out to introduce the functional groups (active sites) on the polymer chain.[9]

The monomers solution (organic phase), also containing a crosslinking agent, a porogen agent and an initiator is added under stirring to a reactor, fitted with a mechanical stirrer, N_2 inlet and a condenser, already containing the aqueous phase (AP), consisting of suspension stabilizers (to avoid bead agglomeration) and a soluble salt (to diminish the monomers solubility in water, by salting out effect), to suspend the organic phase (OP) which, under stirring, forms spherical droplets where the polymerization will occur. The suspension, under stirring, is then heated to promote the polymerization. After polymerization completion, the beads are separated, by filtration, washed to remove stabilizers, porogen agents and unreacted monomers, sieved and finally dried. The bead size is controlled mainly by stirring speed, AP/OP ratio and suspension stabilizers. High stirring speeds and high AP/OP ratios generally produce a smaller bead size. The crosslinking degree of the polymer beads is controlled by varying the proportion of crosslinker in the monomer solution. The beads porosity is determined mainly by the amount and type of porogen system and by the proportion of crosslinker. The porogens can be solvating diluents, non-solvating diluents, soluble polymers or mixtures thereof. The produced beads will not present pores in dry state if the proportion of crosslinker is low and the porogen is a good solvating agent for the polymer chains, in this case the produced beads are gel-type, with no pore structure. At high proportion of crosslinker porous beads can be produced if the porogen is adequately chosen. The pore size varies from less than 20 Å (micropores) to more than 500 Å (macropores). Solvating diluents can produce beads with low pore volume (up to about 0.8 mL/g) and high surface areas $(50 - 500 m^2/g)$, non-solvating diluents tend to produce beads with high pore volumes $(0.6 - 2.0$ mL/g) and surface areas varying from 10 to 100 m^2/g and soluble polymers can result in beads with a pore volume up to 0.5 mL/g, low surface areas $(0 - 10 m^2/g)$ and large average pore diameter.[9,10] Table 1 shows the relation between pore size and porosity.

Table 1: Relation of porosity to pore size

Gel porosity	Pore diameter (Å)		
	Macroporosity	**Mesoporosity**	**Microporosity**
No pores	D > 500	20 < D < 500	D < 20
in dry state	Macropores	Mesopores	Micropores

Mechanism of pore structure formation

The mechanism of pore formation was described by Kun and Kunin [11] as a process in three steps. During the early stages of the polymerization, a polymer composed of linear chains with pendant vinyl groups is formed. As the polymerization progresses intramolecularly crosslinked microgels and higher molecular weight linear chains soluble in the monomer/diluent mixture are produced. The phase separation occurs and two phases are formed in the polymerizing droplets: a copolymer-rich phase and a diluent phase with a very low polymer concentration, the monomers being distributed between the two phases. Since solvated and very lightly reticulated polymers can behave as a liquid the interfacial tension on polymer-rich phase gives it the low-energy spherical form, and the polymer separates as a mass of spherical microgels (microspheres). At a determined value of conversion of monomer to polymer macrogelation occurs and are produced gel-type beads formed by the agglomeration of microgels (microspheres), this is the first step of the pore structure formation. In the second step, the microspheres are bound together through the polymerization of the monomer molecules solvating the polymer chains. It is during this stage that the porous structure is actually formed.[11] The third step involves the polymerization of residual monomer that is completed by increasing the polymerization temperature. The mechanism of pore structure formation is summarized in Figure 1.

1st STAGE

> ♣ microgel formation
> ♣ phase separation (sperical microgels)
> ♣ microgels agglomeration (macrogelation)
> **microspheres**

2nd STAGE

> ♣ microspheres are bound together
> (solvating monomer polymerization)
> **pore structure**

3rd STAGE

> ♣ residual monomer polymerization
> ♣ elimination of porogen agent
> **macroporous polymer beads**

Figure 1: mechanism of pore structure formation

Morphological features of porous polymer beads

Porous polymer beads are constituted of agglomerates (clusters) of microgel spherical particles glued together at their interfaces. Figure 2 presents electron micrographs of two porous poly(styrene-co-divinylbenzene) beads with different porosities.[12] These micrographs reveal the clustered structure and the difference in the size of the agglomerates and pores of the two beads. The visual appearance of those polymer beads is determined by the size of these clusters, and the beads pores are the voids within and between the clusters and the holes between the microspheres from which the agglomerates are formed. Thus, as it was already mentioned in the mechanism of pore structure formation, the porous beads are a two phase material – the gel phase (clusters of microspheres) and the pore phase.[7]

 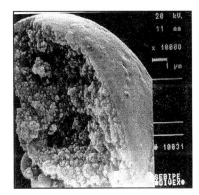

Figure 2: Electron micrographs of two different porous poly(styrene-co-divinylbenzene) beads[12]

When the diameter of the beads internal clusters is approximately 2000 Å or larger, the beads are opaque, when that diameter is smaller than 2000 Å, the beads are translucent. The gel-type beads with no internal pore structure (no pores in dry state) appear transparent (Figure 3).[7]

(a)   (b)

(c)

Figure 3: Optical micrographs of beads with different morphologies: (a) transparent gel-type beads, (b) translucent porous beads, (c) opaque porous beads

Styrene-divinylbenzene copolymers as packing materials for SEC

In general, porous polymers are characterized by a relatively broad pore size distribution. All pores in such materials are typically interconnected and the access to the different pores is controlled by the hydrodynamic volume of the dissolved molecules. The molecules will only penetrate those pores that are able to accommodate their size, while smaller pores remain inaccessible for steric reasons. The principle of size exclusion has been adopted in numerous processes from sieving to advanced membrane technologies. This concept is the basis of the size exclusion chromatography (SEC).[6]

The first report on gel filtration, now denominated size exclusion chromatography, was published in 1959.[13] Further improvement in SEC happened in 1964 with the introduction of macroporous beads of poly(styrene-co-divinylbenzne) by Moore.[14] Those beads were synthesized by aqueous suspension polymerization, a process that produces polymer particles with diameters in the range of 50 to 500 μm and a broad particle size distribution. However, to carry out high performance SEC small copolymer beads with sizes in the range of 5 to 10 μm and with narrow size distribution are required.

Our group (IMA/UFRJ) synthesized styrene-divinylbenzene copolymer beads with sizes in the range from 5 to 10 μm by two different methods: modified suspension polymerization (MSP) and single-step swelling and polymerization (SSWP) (Figure 4).

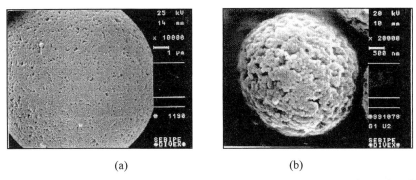

(a)   (b)

Figure 4 : Electron micrographs of poly(styrene-co-divinylbenzene) beads produced by (a) MSP and (b) SSWP [12]

The MSP method, developed by Tanaka and Takeda[15] is a modification of conventional suspension polymerization (CSP). In that method the polymerization is preceded by a high speed stirring (20,500 rpm) dispersion of the organic phase in the aqueous phase at low temperature, to avoid premature polymerization, for a specified time with the objective of obtaining small droplets with sizes around 10 μm. Afterwards, these droplets are transformed in polymer beads, with sizes in that same range, as the polymerization takes place as a CSP.[16,17]

The SSWP method is a two stage technique.[18] In the first stage polystyrene seeds with narrow size distribution (Figure 5) are synthesized by dispersion polymerization in an organic medium. The produced seeds with sizes in the range of 1 to 2 μ are swollen with an emulsified mixture of monomers (styrene and divinylbenzene), initiator and diluent and then submitted to the suspension polymerization conditions to produce beads in the desired size range and with narrow size distribution. An schematic representation of the porous beads synthesis by SSWP method is shown in Figure 6.[12,18]

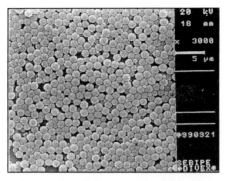

Figure 5: Electron micrograph of polystyrene seeds produced by dispersion polymerization [12]

Poly(styrene-co-divinylbenzene) beads synthesized by SSWP method

The first stage of the SSWP method to produce poly(styrene-co-divinylbenzene) beads is the synthesis of polystyrene seeds by dispersion polymerization initiated by azobisisobutyronitrile (AIBN) in ethanol stabilized by polyvinylpyrrolidone (PVP) of high molecular weight.[19,20] Table 2 shows three conditions employed to prepare polystyrene seeds and their respective characteristics. Figure 6 shows the schematic representation of the SSWP method to produce poly(styrene-co-divinybenzene) beads with small sizes.

Table 2: Polymerization conditions and characteristics of polystyrene seeds[a] [12]

Seeds code	Styrene[b] % V/V	Styrene mL	Ethanol mL	PVP[c] g	Seed diameter μm	Span	M_n x 10^{-4}	M_w x 10^{-5}
PS-34	10	15	135	9	0.94	0.54	6.6	3.44
PS-42	20	60	240	18	1.10	0.62	2.4	0.98
PS-10[d]	20	30	120	1.8	1.73	1.40	2.2	1.46

[a]Initiator = AIBN; [AIBN] = 0.63 %molar (relative to styrene); polymerization temperature = 70°C; stirring speed = 50rpm; polymerization time = 24 h. [b]%Volume relative to (styrene + ethanol). [c] PVP = polyvinylpyrrolidone; M_w = 1.3x10^6; concentration = 60g/L (styrene + ethanol). [d]PVP (K-30), M_w=4.0x10^4, concentration = 60g/L styrene

In the second stage the polystyrene seeds are swollen with an emulsion of the organic phase solution (styrene, divinylbenzene, initiator [2,2'-azobis(2,4-dimethylvaleronitrile) – ADVN], toluene, isoamyl alcohol as compatibilizer and poly(vinyl alcohol)) in water containing poly(vinyl alcohol) (stabilizer) and sodium lauryl sulphate (emulsifier). The polystyrene seeds play two different roles in SSWP method, as templates for the copolymer beads and as porogen agent. The ratio between the amounts of organic phase and seeds influences directly the beads porosity. The lower is that ratio the higher will be the porosity.

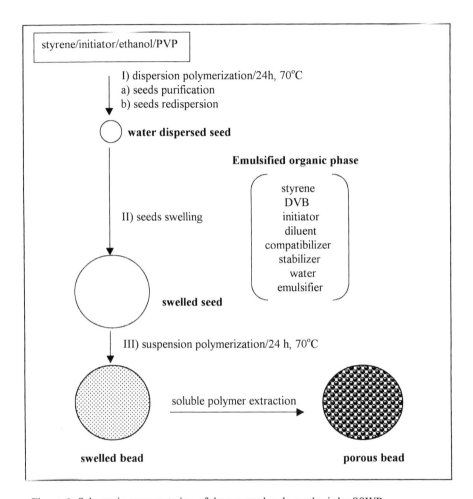

Figure 6: Schematic representation of the porous beads synthesis by SSWP

Chloroform slurries of porous beads were packed, as described previously [21], into stainless-steel columns (30 cm x 7.8 mm i.d.) by a high pressure method to produce SEC columns. The SEC calibration curves were obtained for polystyrene standards using chloroform as eluent at a flow rate of 1.0 mL/min, utilizing a Waters pump (Model 510) equipped with a Waters refractive index detector (Model 410). The upper limit of the range of molecular weight separation (exclusion limit – E.L.) of each column was determined from the calibration curve. The porous structure of the beads define the limits of the separation range and the columns resolution, and the particle size and porosity determine the number of theoretical plates (NTP) also known as efficiency of the column. The average pore diameter (D_i) of the beads in the swelling state in a good solvent is related to the upper limit of the range of molecular weight separation (E.L.) by the equation of Halász[22] ($D_i = 0.62xM^{0.59}$).

Table 3 shows the conditions to produce, from polystyrene seeds, beads of poly(styrene-co-divinylbenzene) with sizes and distribution of sizes adequate to be applied as packing material in size exclusion chromatography. That table also presents the SEC parameters of the columns packed with those materials.[12,18]

Table 3: SSWP conditions and particle size of poly(styrene-co-divinylbenzene) beads[11]

Seeds code	Copolymer code	OP/seeds w/w	Seed diameter μm	Bead diameter μm	E.L.	NTP	R	D_i Å
PS-34	SSW-07	40	0.94	4.49	$> 10^6$	10900	1.1	2150
PS-34	SSW-08	20	0.94	3.55	$> 10^6$	6100	1.6	2150
PS-10	SSW-10	100	1.73	5.03	10^5	6000	1.7	553
PS-42	SSW-11	100	1.10	3.06	10^5	7500	1.7	553

E.L. = exclusion limit; NTP = number of theoretical plates; R = resolution; D_i = average pore diameter in the swelling state, calculated by the equation $D_i = 0.62xM^{0.59}$, where M is the weight average molecular weight.[22] DVB = divinylbenzene; Sty = styrene; DVB/Sty = 65/35 (molar ratio); toluene/(DVB + Sty) (w/w) = 1/1; isoamyl alcohol = 5% w/w (relative to organic phase); organic phase (OP) = DVB + Sty + toluene; polymerization temperature = 70°C; polymerization time = 10h; stirring speed = 100rpm; initiator = ADVN; [ADVN] = 1% molar (relative to the monomers); aqueous phase = AP; OP/AP (w/w) = 1/10; AP = water (ten times the organic phase weight) + 0.25% (w/w) of sodium lauryl sulphate + 0.6% (w/w) of poly(vinyl alcohol).

The beads produced with the seeds PS-34 using the lowest OP/seeds ratio (40 and 20) (SSW-07 and SSW-08) (Table 3) presented high porosity, that was evidenced by SEM technique (Figure 7-a and 8-a) and by the high values of exclusion limit ($>10^6$).

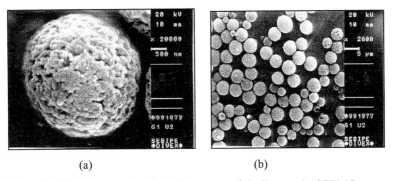

(a)   (b)

Figure 7: SEM micrographs of poly(styrene-co-divinylbenzene) –SSW-07

However, SEM micrographs of those beads with lower magnifications (Figures 7-b and 8-b) showed that non-porous beads were also formed. The non-porous beads formation may be attributed to the polymerization of organic droplets that were not absorbed by the polystyrene seeds. As the diluent of the organic phase was toluene, a good solvent for the produced copolymer, non-porous beads, by conventional suspension polymerization (CSP), were formed out of the seeds, the true reactor in SSWP method. The non-porous beads are responsible for the low values of resolution (1.1 and 1.6) of the SEC columns, inferior to the minimum value (1.8) required for an efficient column. The NTP's, which minimum value is around 10000, were also affected by the presence of non-porous beads.

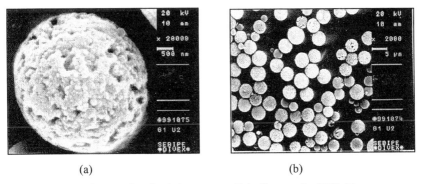

(a)   (b)

Figure 8: SEM micrographs of poly(styrene-co-divinylbenzene) –SSW-08

The beads synthesized with the highest (100) OP/seeds ratio (SSW-10 and SSW-11), i.e. the lowest proportion of polystyrene seeds, presented the lowest porosity (Figure 9) and consequently the lowest exclusion limit. This behavior is in accordance with the seeds'role as porogen agent.

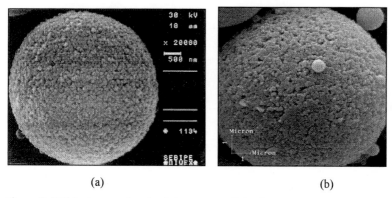

(a)   (b)

Figure 9: SEM micrographs of poly(styrene-co-divinylbenzene) – (a) SSW-10; (b) SSW-11

Poly(styrene-co-divinylbenzene) beads synthesized by MSP method

The second approach to the synthesis of porous copolymer beads with sizes adequate for application as SEC columns packing materials is the MSP. This suspension polymerization method is preceded by the dispersion of organic phase in the aqueous phase at temperature not exceeding 10°C at a stirring speed of 20,500 rpm during 10 minutes. Afterwards the stirring is slow down to 500 rpm, the reaction mixture is heated to 70°C and maintained at this temperature during the polymerization period (10 hours).[16,17]

Poly(styrene-co-divinylbenzene) beads were prepared by MSP method employing two different systems as porogen agent: a toluene/heptane (30/70 v/v) mixture and a toluene solution of polystyrene (PS-10; M_w = 146000; M_n = 22000). In the system using toluene/heptane as porogen only the molar ratio DVB/Sty was varied. The average bead diameter became around 10 μm (Table 4). The SEM micrographs of the beads (Figure 10) show that the beads'porosity increased as the DVB content increased. That behavior is in accordance with the mechanism of porosity formation. Higher amounts of DVB (crosslinking agent) provokes the earlier occurrence of macrogelation and consequent phase separation responsible for the pores formation. As expected the SEC columns exclusion limit increased as the DVB proportion increased from 55% to 75%, but did not vary with the posterior increase of DVB from 75% to 100%. This behavior needs to be further investigated because SM-13 (100% DVB) is accentuately more porous than SM-

39

12 (75% DVB). Probably, this result is due to some problem during the column packing.

Table 4: MSP conditions and particle size of poly(styrene-co-divinylbenzene) beads obtained using as porogen a 30/70 toluene/heptane mixture[a][12]

Copolymer code	DVB/Sty molar ratio	Bead diameter μm	E.L.	NTP	R	D_i Å
SM-09	55/45	9.34	1.5×10^5	4000	1.1	702
SM-12	75/25	10.55	4.22×10^5	3800	3.8	1292
SM-13	100/0	11.59	4.22×10^5	3700	1.7	1292

E.L. = exclusion limit; NTP = number of theoretical plates; R = resolution; D_i = average pore diameter in the swelling state, calculated by the equation $D_i = 0.62 \times M^{0.59}$, where M is the weight average molecular weight.[22] [a]Polymerization initiator = ADVN; [ADVN] = 1%molar (relative to the monomers); pre-stirring time at 10°C = 10 min; polymerization time = 10 h; polymerization temperature = 70°C; porogen = toluene/heptane (30/70 v/v); (toluene + heptane)/(Sty + DVB) = 2/1 (v/v); suspension stabilizer = poly(vinylalcohol) (PVA); aqueous phase (AP) = water + 0.5% w/v PVA; organic phase (OP) = (styrene + divinylbenzene + porogen); AP/OP = 4/1 (v/v); polymerization stirring speed = 500 rpm

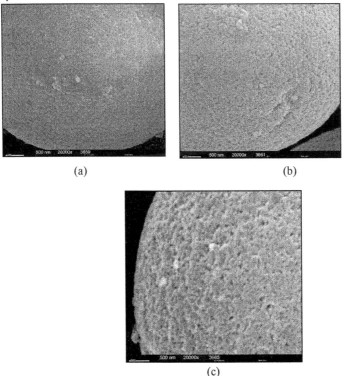

(a) (b)

(c)

Figure 10: SEM micrographs of poly(styrene-co-divinylbenzene) – (a) SM-09; (b) SM-12; (c) SM-13

The MSP method using as porogen agent a toluene solution of polystyrene (PS) produced beads with different porosities as the amount of polystyrene or DVB was varied. Using a 6% PS toluene solution and varying the DVB proportion (55%, 75%, 100%) were obtained beads with average diameter around 11 μm and with porosities dependent on the DVB content (Table 5). The SEM micrographs of the beads (Figure 11) show a slight porosity increase as the DVB content increased. It was observed a tendency to higher values of porosity for these beads than for the ones produced in presence of toluene/heptane mixture as porgen agent. The exclusion limit of the SEC columns increased as the DVB proportion increased from 55 to 100%, but remained constant when the DVB amount was increased from 55 to 75%. Probably this result is due to the small difference in porosity of those two copolymer beads.

Table 5: MSP conditions and particle size of poly(styrene-co-divinylbenzene) beads obtained using as porogen a 6% PS toluene solution[a][12]

Copolymer code	DVB/Sty molar ratio	Bead diameter μm	E.L.	NTP	R	D_i Å
SM-35	55/45	11.1	4.22×10^5	6800	1.1	1292
SM-39	75/25	11.8	4.22×10^5	3600	1.1	1292
SM-37	100/0	11.3	8.00×10^5	4300	1.1	1885

E.L. = exclusion limit; NTP = number of theoretical plates; R = resolution; D_i = average pore diameter in the swelling state, calculated by the equation $D_i = 0.62 \times M^{0.59}$, where M is the weight average molecular weight.[22] [a]Polymerization initiator = ADVN; [ADVN] = 1%molar (relative to the monomers); pre-stirring time at 10°C = 10 min; polymerization time = 10 h; polymerization temperature = 70°C; porogen = 6% PS toluene solution (w/w); toluene/(Sty + DVB) = 1/1 (v/v); suspension stabilizer = poly(vinylalcohol) (PVA); aqueous phase (AP) = water + 0.5% w/v PVA; organic phase (OP) = (styrene + divinylbenzene + toluene); AP/OP = 4/1 (v/v); polymerization stirring speed = 500 rpm

The largest variations of porosity occurred when the amount of DVB was fixed at 100% and the amount of polystyrene in the porogen solution was varied from 1% to 6% (Table 6). This behavior was translated by the accentuated variation of the exclusion limit of the SEC columns obtained with these beads. The beads synthesized with 1% of porogen produced a SEC column which exclusion limit was 1.02×10^4 while the column packed with the beads synthesized with 6% of porogen presented a exclusion limit of 8.0×10^5.

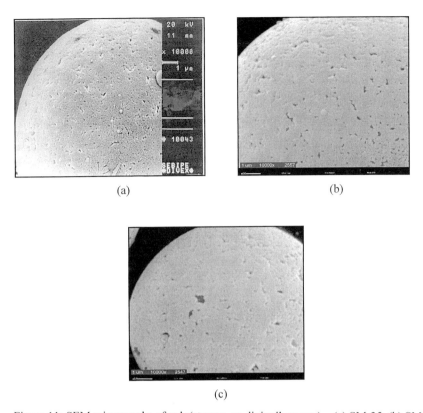

(a)   (b)

(c)

Figure 11: SEM micrographs of poly(styrene-co-divinylbenzene) – (a) SM-35; (b) SM-39; (c) SM-37

Table 6: MSP conditions and particle size of poly(styrene-co-divinylbenzene) beads obtained using as porogen PS toluene solutions at 100% DVB[a][12]

Copolymer code	PS/toluene w/w %	Bead diameter μm	E.L.	NTP	R	D_i Å
SM-23	1	11.1	1.02×10^4	900	1.5	144
SM 24	4	9.9	1.42×10^3	1700	1.0	680
SM-37	6	11.3	8.00×10^5	4300	1.1	1885

E.L. = exclusion limit; NTP = number of theoretical plates; R = resolution; D_i = average pore diameter in the swelling state, calculated by the equation $D_i = 0.62 \times M^{0.59}$, where M is the weight average molecular weight.[22] [a]Polymerization initiator = ADVN; [ADVN] = 1%molar (relative to the monomers); pre-stirring time at 10°C = 10 min; polymerization time = 10 h; polymerization temperature = 70°C; DVB = 100%; toluene/(DVB) = 1/1 (v/v); suspension stabilizer = poly(vinylalcohol) (PVA); aqueous phase (AP) = water + 0.5% w/v PVA; organic phase (OP) = (divinylbenzene + toluene); AP/OP = 4/1 (v/v); polymerization stirring speed = 500 rpm

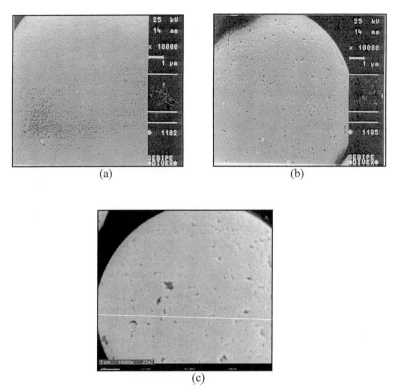

(c)

Figure 12: SEM micrographs of poly(styrene-co-divinylbenzene) – (a) SM-23; (b) SM-24; (c) SM-37

Conclusion

1) The highest value of exclusion limit (10^6) for the SEC columns packed with poly(styrene-co-divinylbenzene) beads was attained when the beads were synthesized, employing polystyrene seeds, by the single-step swelling and polymerization (SSWP) method using high proportion of seeds (OP/seeds = 20 or 40).

2) The poly(styrene-co-divinylbenzene) beads prepared by modified suspension polymerization (MSP) method were, in general, less porous than the beads obtained by SSWP method, in equivalent experimental conditions.

3) The most porous poly(styrene-co-divinylbenzene) beads, prepared by MSP method, were those synthesized using polystyrene/toluene solution as porogen agent. The maximum value of exclusion limit attained, in this case, was 8.0×10^5.

4) High proportions of polystyrene, as porogen agent, and DVB produced the SEC columns with the highest values of exclusion limit.

[1] F.M.B.Coutinho, M.A.F.S.Neves, M.L.Dias, *J.Appl.Polym.Sci.* **1997**, 65, 1257.

[2] P.H.Toy, K.D.Janada, *Tetrahedron Let.* **1999**, 40, 6329.

[3] D.C.Sherrington, *Nouv.J.Chim.* **1982**, 6, 661.

[4] F.M.B.Coutinho, V.G.Teixeira, C.C.R.Barbosa, *J.Appl.Polym.Sci.* **1998**, 67, 781.

[5] M.M.Sharma, *React.Funct.Polym.* **1995**, 26, 3.

[6] V.Smigol, F.Svec, J.M.J.Fréchet, *Macromolecules* **1993**, 26, 5615.

[7] R.L.Albright, *React.Polym* **1986**,4, 155.

[8] A.Akelah, D.C.Sherrington, *Chem.Rev.* **1981**, 81, 557.

[9] R.A.Beauvais, S.D.Alexandratos, *React.Funct.Polym.* **1998**, 36, 113.

[10] W.L.Sederel, G.J.Jong, *J.Appl.Polym.Sci.* **1973**, 17, 2835

[11] K.A.Kun, R.Kunin, *J.Polym.Sci. PartA-1* **1968**, 6, 2689

[12] M.A.F.S.Neves, *DSc Thesis* **2001**, Universidade Federal do Rio de Janeiro, Brasil

[13] J.Porath, P.Flodin, *Nature* **1959**, 183, 1657.

[14] J.C.Moore, *J.Polym.Sci. A2* **1964**, 835

[15] Y.Tanaka, J.Takeda, US Patent 4.338.404 (1982).

[16] M.A.F.S.Neves, *MSc Thesis* **1996**, Universidade Federal do Rio de Janeiro, Brasil

[17] G.M.Alves, *MSc Thesis* **1999**, Universidade Federal do Rio de Janeiro, Brasil

[18] K.Ogino, H.Sato, K.Tsuchiya, H.Suzuki, S.Moriguchi, *J.Chromatogr.A* **1995**, 699, 59.

[19] C.M.Tseng, Y.Y.Lu, M.S.El-Aasser, J.W.Vanderhoff, *J.Polym.Sci. PartA* **1986**, 24, 2995.

[20] A.J.Paine, W.Luymes, J.McNulty, *Macromolecules* **1990**, 23, 2569.

[21] F.M.B.Coutinho, M.A.F.S.Neves, M.L.Dias, *Polímeros: Ciência e Tecnologia* **1999**, 3, 71.

[22] I.Halász, K.Martin, *Angew.Chem.Int.Ed.Engl.* **1978**, 17, 901.

Compatibilization of Elastomer-Based Blends

Bluma G. Soares*, Alex S. Sirqueira, Marcia G. Oliveira,
Mauricio S. M. Almeida

Instituto de Macromoléculas, Universidade Federal do Rio de Janeiro,
Centro de Tecnologia, Bl. J, Ilha do Fundão, 21945-970, Rio de Janeiro, RJ,
Brasil
E-mail: bluma@ima.ufrj.br

Summary: The reactive compatibilization of ethylene-propylene-diene
(EPDM)-based dissimilar elastomer blends has been investigated in terms of
mechanical properties and swelling degree. The use of mercapto-
functionalized copolymers resulted in an improvement of mechanical
properties of natural rubber-EPDM blends. The mercapto-groups are able to
react with the carbon-carbon double bonds of the high diene rubber,
resulting in a good interaction between phases. These interactions were
confirmed by the amount of insoluble material obtained in non-vulcanized
blends. From dynamic mechanical properties and swelling degree, one can
suggest a covulcanization process in these blends cured with sulfur-based
system.

Blends composed by nitrile rubber with EPDM displayed good results in
terms of mechanical properties when mercapto-functionalized EVA was
employed instead of functionalized EPDM, probably because of the higher
polarity of the former associated to its lower viscosity. Additionally, an
improvement on mechanical properties was also achieved by using EPDM
functionalized with mercapto or anhydride groups in combination with
nitrile rubber functionalized with epoxy or oxazoline groups.

Introduction

Blending of two elastomers is carried out for several purposes, such as enhancement of
physical properties, improvement of processing characteristics and ageing resistance as
well as decreasing compound cost. Most of polymer blends are heterogeneous and
incompatible, resulting in a gross-phase separated morphology, poor interfacial
adhesion and consequently poor mechanical properties. In the specific case of dissimilar
elastomer blends, there is also the problem of cure-rate incompatibility because of
differences in unsaturation and polarity of the elastomers. Due to the higher solubility of
sulfur in unsaturated rubber, the curatives tend to diffuse into the elastomer phase in
which their solubility is highest, resulting in a significant difference on the crosslink
density of each rubber phase.[1-3]

CCC 1022-1360/00/$ 17.50+.50/0

Several approaches have been made to develop vulcanized blends with a single network structure including crosslinked macromolecules of both polymers across the macrodomain interfaces. Most of these studies are related to natural rubber (NR)/ ethylene-propylene-diene (EPDM) rubber blends and include: substitution of zinc oxide by lead oxide as a component of the activation system, [4] halogenation of EPDM to improve sulfur vulcanization, [5] chemical modification of the accelerators to improve their solubility in both phases, [6] grafting of the accelerator onto the phase that present lower solubility with the accelerators, such as EPDM, [7] pre-curing the slower curing rubber (EPDM)[8] and the use of specific accelerators which can react with the EPDM phase, making it more polar and compatible with the curing agents, together with a two-stage vulcanization.[9]

The functionalization of EPDM with maleic anhydride has been also employed by Coran to improve cure compatibility of NR/EPDM blends.[10-11] This functionalization permits the EPDM to be crosslinked independently with the zinc oxide in the accelerated- sulfur vulcanizing system. More recently, the use of poly-trans-octylene (TOR) as compatibilizing agent in NR/EPDM blends has been reported to produce a fine morphology and improve the properties of these blends. [12] TOR acts at the interface between the blend components because of its low viscosity.

The compatibilization and covulcanization of nitrile rubber (NBR)/EPDM blends are even more difficult because the components are different in unsaturation level and polarity. Some reports employ TOR, [13] polybutadiene, [14] polychloroprene, [14-17] chlorinated polyethylene, [14,18] and chlorosulfonated polyethylene. [14,18] All these compounds increased the compatibility of NBR/EPDM blends, although the authors have considered the chlorinated polyethylene as the universal compatibilizer for NBR/EPDM blends. [14,18]

The reactive compatibilization in elastomer blends

Based on the technological interest of several rubber industries on elastomer blends with good mechanical performance and other interesting properties, we decided to develop new systems able to improve the interfacial adhesion between elastomer phases and, in some cases, impart as much as possible a good crosslink distribution between the phases. Our goal is to incorporate low-unsaturated rubbers (such as EPDM) to high-unsaturated elastomer matrix in order to achieve elastomer materials with outstanding resistance to the action of heat, oxygen and ozone. For this purpose, we have focused our studies on reactive compatibilization.

The reactive compatibilization is based on the addition of a functionalized polymer that reacts with one or both blend components, thus promoting an effective anchorage between the phases, through chemical linkage.

The first strategy used in our researches to achieve the compatibilization in elastomer-based blends involves the ability of mercapto groups in reacting with carbon-carbon double bonds of unsaturated polymers.[19-21] Therefore, mercapto groups have been introduced along the backbone of the low-unsaturated (such as EPDM) [22] or saturated (such as ethylene- vinyl acetate EVA copolymer) [23] blend component using simple functionalization reactions, as illustrated in Figure 1. An effective interaction between the blend components can be achieved with the addition of small amount of these copolymers, since the mercapto groups react with the double bonds of high-diene rubbers whereas the main chain of the reactive compatibilizing agent may physically interact with the other component, similar in nature. This approach has been profitable for the compatibilization of EVA-based blends, such as NR/EVA,[23-25] NBR/EVA,[26] and SBR/EVA.[27] As it will be discussed in next section, EPDM-based blends have also shown an improvement on the mechanical performance with this kind of compatibilization.

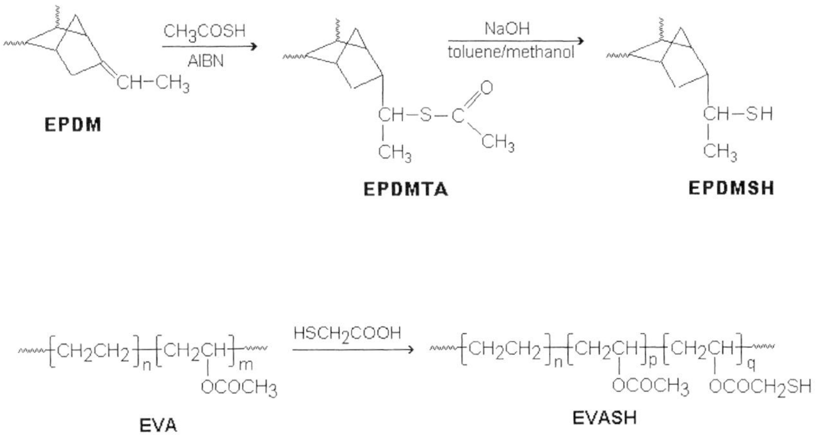

Figure 1. Scheme for the mercapto-functionalization of EPDM and EVA

The other strategy involves two different reactive polymers, whose functional groups react each other. A small portion of these compounds are separately blended with each elastomer component and the resulting masterbatches are then blended together to form a compatible blend, whose phases are bounded through a new functional group formed during the mixing process. Some functionalized polymers used in this work are illustrated in Table 1. This procedure was employed in NBR/EPDM because of the possibility of introducing reactive groups along the NBR backbone.

Table 1. Examples of reactive compatibilization between elastomer blends

High-unsaturated rubber	Low-unsaturated rubber	Functional group formed during mixing process
NBR	EVA or EPDM	$-\overset{\overset{O}{\|\|}}{C}-NH\text{-}CH_2\text{-}CH_2\text{-}S-$
	EPDM	$-\overset{\overset{O}{\|\|}}{C}-NH\text{-}CH_2\text{-}CH_2\text{-}O-\overset{\overset{O}{\|\|}}{C}$ ... COOH
NBR	EVA or EPDM	$-CH-S-$; $CH_2\text{-}OH$
	EPDM	$-CH-O-\overset{\overset{O}{\|\|}}{C}$ ... COOH ; $CH_2\text{-}OH$

The compatibilization of NR/EPDM blends

The NR/EPDM blends have been compatibilized with the addition of 2.5 phr of mercapto-modified EPDM (EPDMSH). The curing characteristics and mechanical properties are summarized in Table 2. The first indication of covulcanization was obtained from swelling results. In this Table, Vr corresponds to the volume fraction of the rubber in a swollen network. Higher Vr values mean lower swelling degree. According to Shershnev,[28] if interfacial bonds are formed during covulcanization, the lightly swollen dispersed phase in a blend will restrict the swelling of the swollen continuous phase. Such interfacial bonds must have been also formed with the addition of EPDMSH in NR/EPDM blends, since a substantial decrease of swelling degree was observed with the compatibilization. This phenomenon is also confirmed by the amount of insoluble material isolated after submitting the vulcanized samples to a extraction with xylene at 120°C for 24h. The amount of insoluble material is higher than the amount of the NR component in blends containing 70 and 60 wt% of NR, indicating that EPDMSH also promotes some crosslinking into the EPDM phase. The compatibilizing effect of EPDMSH can be confirmed from the results concerning ultimate tensile strength. As observed in Table 2, the addition of 2.5 phr of EPDMSH increases this property without affecting significantly the elongation at break.

Table 2. Curing parameters and mechanical properties of NR/EPDM blends as a function of composition and compatibilization. [a]

Properties	NR/EPDM (wt%)					
	80:20		70:30		60:40	
	b	c	b	c	b	c
M_L (lb.in) [d]	1.5	1.9	1.5	2.5	2.0	2.0
M_H (lb.in) [d]	16.5	17.0	19.0	20.3	19.5	20.0
t_{s1} (min) [e]	3.0	2.8	2.8	2.3	3.3	3.0
t_{90} (min) [e]	8.0	8.0	9.5	7.4	10.3	9.5
Vr [f]	0.14	0.20	0.12	0.19	0.12	0.17
Gel content (%)	74	73	66	80	74	80
σ_B (Mpa) [g]	10.3	13.0	10.5	13.5	9.4	15.0
ε_B (%) [h]	850	830	900	840	600	700

a) The blends were prepared in a two roll mill operating at 80°C and at 20 rpm. NR was masticated for 2 min and then EPDM and the functionalized compatibilizing (EPDMSH) were subsequently added. After the homogenization of the rubber blend (at about 4 min), the other ingredients were added in the following order: zinc oxide (5.0 phr), stearic acid (1.5 phr), Irganox 245 (1.0 phr), sulfur (2.5 phr) and 2,2'-dithiobisbenzothiazole (MBTS) (0.8 phr). The processing time after each component addition was about 2 min
b) Blends without compatibilizer
c) Blends with 2.5 phr of EPDMSH
d) M_L and M_H are the minimum and maximum torque obtained from oscillating disk rheometer, respectively.
e) t_{s1} and t_{90} are the scorch time and optimum curing time, respectively.
f) Vr corresponds to the volume fraction of the rubber in a swollen network and is calculated as follows: $Vr = \dfrac{D \times \rho_r^{-1}}{D \times \rho_r^{-1} + (S-D) \times \rho_s^{-1}}$
where: D = deswollen weight, S = swollen weight, ρ_r = density of rubber blend and ρ_s = density of solvent
g) Ultimate tensile strength.
h) Elongation at break.

50

In addition of a compatibilizing effect, this functionalized copolymer has also favored the crosslink inside the EPDM phase. This phenomenon was suggested from dynamic mechanical analysis. Figure 2 compares the damping (tan δ) of the vulcanized components with those of the 70:30 NR/EPDM blends, evaluated from -80 to 25⁰C. The non-compatibilized blend displays only one transition at a temperature below to that one found in pure NR sample, which can be attributed to the vulcanized NR phase together with the non vulcanized EPDM phase whose transition should occur at very similar temperatures. The presence of the non vulcanized EPDM fraction in the blend may also be responsible for the lower glass transition temperature of this blend when compared to the pure NR component.

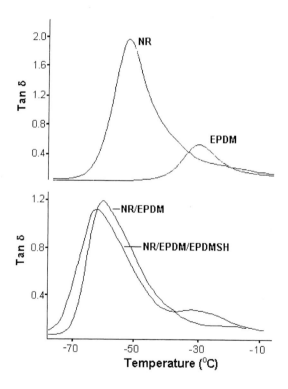

Figure 2. Dependence of tan δ with the temperature for vulcanizate samples of NR, EPDM and NR/EPDM (70:30) and NR/EPDM/EPDMSH (70:30:2.5)

NR/EPDM blend compatibilized with EPDMSH presents a lower value of the glass transition temperature related to the NR phase and also a lower damping. These results may also be attributed to the effective compatibilization of this functionalized copolymer. The strong interactions between the mercapto groups along the EPDMSH backbone and the NR phase decrease the mobility of this phase, giving rise to a decreasing of the damping. It is important to emphasize the presence of a second transition in this blend, at a temperature similar to that found for the vulcanized EPDM sample. This second transition at higher temperature may be related to the proportion of vulcanized EPDM phase in the NR/EPDM blend compatibilized with EPDMSH. These results are in agreement to those concerning gel content.

In order to analyze the effectiveness of the reactive compatibilization, we have studied non-vulcanized NR/EPDM blend containing 10 phr of the functionalized copolymer, as compatibilizing agent. The compatibilized and noncompatibilized NR/EPDM (50:50 %) blends were prepared by melt mixing the components in a two-roll mill at 80^0C for 15 min and compression-molded at 160^0 for 10 min in a hydraulic press at 6.7 MPa. The samples were submitted to extraction with hot xylene for 24 hours in order to determine the amount of insoluble material. The noncompatibilized blend did not present any insoluble material, as expected since there is no vulcanizing system in the formulation and no reaction is expected to occur between NR and EPDM during the blend processing.

The presence of EPDMSH resulted in a considerable amount of insoluble material (around 40%). This proportion is higher than the amount of the functionalized copolymer used in the blend, suggesting the formation of a network during the process.

The composition of the insoluble fraction obtained from the compatibilized blend was determined by thermogravimetric analysis. Figure 3 compares the Thermogravimetric curves of the insoluble fraction with those of the pure components. As expect, the degradation of NR occurs at a lower temperature than EPDM. The insoluble material obtained from the compatibilized blend (Fig.3c) presents two degradation steps, which can be related to the NR and EPDM phases. Since the amount of EPDM in this insoluble fraction is higher than the amount of the functionalized copolymer added in the blend, we can conclude that the reactive compatibilization involves the reaction of the mercapto groups of the EPDMSH with both NR and EPDM phases in the blend. In other words, the functionalized copolymer is acting as a reactive interfacial agent.

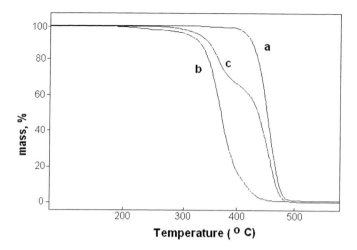

Figure 3. Thermogravimetric analysis of (a) EPDM, (b) NR and (c) the insoluble material obtained from non vulcanized NR/EPDM (50:50 wt%) blend, compatibilized with 10 phr of EPDMSH

The compatibilization of NBR/EPDM blends

NBR/EPDM blends are highly incompatible because of the difference in polarity and unsaturation level between the components. Besides the poor interfacial adhesion and gross-phase separated morphology originated from very different structures, the cure-rate incompatibility is even more significant, specially when sulfur-based curing systems are employed because of the great solubility of the curatives in the polar and unsaturated NBR phase.

In this preliminary study, several reactive compatibilizing systems have been employed in NBR/EPDM (70:30 wt%) blends. Table 3 presents the tensile properties of vulcanized blends containing mercapto-functionalized polymers (EPDMSH or EVASH). A commercial sample of EPDM functionalized with maleic anhydride (EPDM-MA) was also employed in order to compare the compatibilizing efficiency.

Table 3. Mechanical properties of vulcanized NBR/EPDM (70:30 wt%) blends, compatibilized with mercapto-functionalized copolymers. [a]

Compatibilizing agent	Vr	Insoluble material [b] (%)	Mechanical properties			
			Ultimate tensile strength (MPa)		Elongation at break (%)	
			c	d	c	d
None	0.20	0	2.7	2.4	600	560
EPDMSH	0.19	10	2.3	2.7	580	630
EPDM-MA	0.18	26	3.2	4.0	660	690
EVA18SH [e]	0.20	8	4.2	5.6	750	750
EVA28SH [f]	0.18	5	4.3	4.4	730	670

a) The blends were prepared in a two roll mill operating at 110°C. NBR was masticated for 2 min and then EPDM and the functionalized compatibilizing were subsequently added. After the homogenization of the rubber blend (at about 4 min), the other ingredients were added in the following order: zinc oxide (5.0 phr), stearic acid (0.5 phr), sulfur (1.0 phr) and 2,2'-dithiobisbenzothiazole (MBTS) (2.0 phr).
b) Insoluble material obtained from non vulcanized blends (before the addition of curatives).
c) Before ageing
d) After ageing in an air circullating oven at 70°C, for 72h.
e) EVA18SH= EVA with 18% of vinyl acetate functionalized with SH.
f) EVA28SH = EVA with 28% of vinyl acetate functionalized with SH.

An increase of tensile properties has been observed with the addition of 5 phr of EPDM-MA or EVASH. EPDMSH does not improve these properties, in spite of being constituted with the same backbone of the EPDM phase and having mercapto groups able to react with the NBR phase.

In order to evaluate the extent of chemical reaction in the presence of these reactive copolymers, a small sample of each blend was withdrawn from the roll-mill before the addition of curatives and submitted to extraction with hot toluene for 24h. As observed in Table 3, all funcionalized copolymers produced insoluble materials, indicating a reactive interaction between the phases. The lower amount of insoluble material obtained with EVA18SH or EVA28SH is due to the low amount of mercapto groups in these samples.

The best results achieved with EVASH may be attributed to its lower melt viscosity and intermediary surface tension. Both characteristics tend to locate this component at the interface. This morphological situation should enhance the interaction of the mercapto groups with the double bonds of both rubber phases, NBR and EPDM. EPDM-MA also improves the tensile properties probably because the anhydride groups interact with zinc

54

oxide, bringing some amount of the curatives inside the EPDM phase.

Considering that the swelling degree was not affected by the addition of the functionalized copolymers, one can assume that the good mechanical performance may be better attributed to some interfacial action and not to the covulcanization phenomenon.

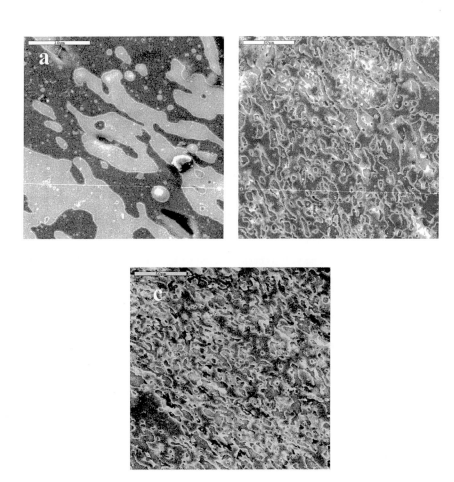

Figure 4. SEM micrographs of non vulcanized NBR/EPDM (70:30 wt%) blends: (a) non compatibilized blend; (b) compatibilized with 5 phr of EVA18SH and (c) compatibilized with 5 phr of EVA28SH.

The morphologies of NBR/EPDM blends compatibilized with EVASH are compared to that of non-compatibilized one in Figure 4. These micrographys were taken from non-vulcanized blends, in order to observe the effect of the EVASH as compatibilizing agent without the influence of the curatives. The samples were taken from the roll-mill before the addition of curatives, pressed at $160^{0}C$ for 15min and analyzed in a ZEISS scanning electron microscope using backscattered detector. The sample surfaces were first stained with osmium tetroxide. The light region in the micrographs presented in Figure 4 corresponds to the NBR phase stained with osmium tetroxide and the black region corresponds to the EPDM phase. The noncompatibilized NBR/EPDM (70:30 wt%) blend presents a gross-phase separated morphology (Fig 4a), whereas the blends compatibilized with 5 phr of EVA18SH (Fig 4b) or EVA28SH (Fig 4c) present a substantial decrease of the phase size, confirming the interfacial action.

In order to increase the interaction between the phases, the NBR component was also functionalized with epoxy groups (NBR-epo) or oxazoline groups (NBR-ox), because these groups present higher reactivity towards mercapto and carboxyl groups than carbon-carbon double bonds. These reactions were summarized in Table 1. For this study, masterbatches of NBR/NBRepo or NBR/NBRox in a proportion of 100:5 phr and EPDM/EPDM functionalized in a proportion of 100:10 phr were separately prepared. Then, 70 parts of the NBR masterbatch and 30 parts of EPDM masterbatch were blended together before adding the curatives. The results are summarized in Table 4. This new procedure resulted in blends with better tensile properties than those without the presence of antagonic reactive groups in the NBR phase. The best results were again achieved with EVASH. Concerning the antagonic reactive group inside the NBR phase, the oxazoline group was more effective on the reactive compatibilization.

Most of the compatibilized blends presented an increase of the ultimate tensile strength after ageing in an air-circulating oven for 72h, indicating a post-curing process during ageing. This phenomenon is accomplished with a small decrease of the elongation at break.

The reaction between the functional groups inside the rubber phases can be suggested from extraction experiments in hot toluene. These experiments were carried out in non vulcanized blends, obtained just before the addition of the curatives. As indicated in Table 4, the amount of insoluble material was significant in all reactive blends. The highest proportion has been achieved in systems containing NBR functionalized with

56

oxazoline (NBRox), because this group presents higher reactivity than epoxy groups towards mercapto or carboxyl groups located inside the EPDM phase. It is important to emphasize that the amount of these insoluble material is higher than the amount of the functionalized copolymer used in the masterbatches. In NBR masterbatch, the proportion of NBRepo or NBRox corresponds to 5%. Since 70% of these compounded rubbers were used in the blend, the amount of functionalized material corresponds to around 3.7%. On the same way, the EPDM masterbatch contains 10% of the functionalized material, which corresponds to 3% of the functionalized copolymer in the blend. Therefore, one can suggest that the insoluble material also includes some amount of non functionalized rubber, which interacts with the functionalized copolymers through their double bonds.

Conclusions

Dissimilar elastomer blends can be compatibilized with functionalized copolymers. The components in NR/EPDM blends present different unsaturation level but are similar in polarity. In this case, the use of small amount of mercapto-functionalized EPDM was beneficial for the improvement of mechanical performance. From swelling experiments and dynamic mechanical properties, one can suggest that a covulcanization process occurs in some extent together with the reactive compatibilization.

NBR/EPDM blends are more difficult to be compatibilized because of the differences in both unsaturation level and polarity. Good mechanical performance was achieved only with EVA functionalized with mercapto groups (EVASH), probably because of its location at the interface as a consequence of its lower viscosity and intermediary polarity. This compound resulted in blends with uniform morphology and smaller phase size.

One of the best way to develop NBR/EPDM blends with outstanding properties (good mechanical performance and good ageing resistance) consists of employing antagonic functional groups inside each rubber phase. When part of the NBR phase is functionalized with epoxy or oxazoline groups, the reaction with EPDM functionalized with mercapto or anhydride groups is more effective. With this new system, the use of NBR functionalized with oxazoline groups is much better due to its higher reactivity towards mercapto or carboxyl groups.

Table 4. Mechanical properties of vulcanized NBR/EPDM (70:30 wt%) blends, compatibilized with a combination of two functionalized copolymers. [a]

Blend components		Vr	b (%)	Ultimate tensile strength (MPa)		Elongation at break (%)	
				c	d	c	d
NBR	EPDM	0.20	0	2.7	2.4	600	560
NBR/NBRepo	EPDM/EPDMMA	0.21	17	3.6	4.2	650	620
NBR/NBRepo	EPDM/EPDMSH	0.19	12	2.9	3.2	610	590
NBR/NBRepo	EPDM/EVA18SH	0.21	10	5.2	4.2	750	700
NBR/NBRox	EPDM/EPDM-MA	0.21	23	4.9	4.9	660	630
NBR/NBRox	EPDM/EPDMSH	0.21	26	3.3	4.0	610	610
NBR/NBRox	EPDM/EVA18SH	0.21	14	6.0	6.4	760	730

a) The blends were prepared in a two roll mill operating at 110°C. NBR masterbath was blended with EPDM masterbatch. After the homogenization of the rubber blend (at about 4 min), the other ingredients were added in the following order: zinc oxide (5.0 phr), stearic acid (0.5 phr), sulfur (1.0 phr) and 2,2'-dithiobisbenzothiazole (MBTS) (2.0 phr).
b) Insoluble material obtained from non vulcanized blends (before the addition of curatives).
c) Before ageing
d) After ageing in an air-circulating oven at 70°C, for 72h.

Aknowledgements

We would like to acknowledge the Conselho Nacional de Desenvolvimento Científico e Tecnológico - CNPq, Coordenação de Aperfeiçoamento de Pessoal de Ensino Superior - CAPES, CEPG-UFRJ, FAPERJ and PADCT/CNPq (proc. nº 620132/98-1) for the financial support and to NITRIFLEX S. A. and Michelin do Brasil for the polymer supplying.

References

[1] C.M. Roland, Rubber Chem. Technol. **1989**,62, 456.
[2] G. J. Van Amerongen, Rubber Chem. Technol. **1964**,37, 1065.
[3] J. R. Gardiner, Rubber Chem. Technol. **1968**,41, 1312.
[4] M. E. Woods, J. A. Davidson, Rubber Chem. Technol. **1976**, 49, 112.
[5] R. T. Morrissey, Rubber Chem. Technol. **1971**, 44, 1025.
[6] R. P. Mastromatteo, J. M. Mitchell, T. J. Brett Jr, Rubber Chem. Technol. **1971**, 44, 1065.

[7] K. C. Baranwall, P. N. Son, Rubber Chem. Technol.**1978,** 47, 88.

[8] N. Suma, R. Joseph, K. E. George, J. Appl. Polym. Sci, **1993,** 49, 549.

[9] A. K. Ghosh, S. C. Debnath, N. Naskar, D. K. Basu, J. Appl. Polym. Sci. **2001,** 81, 800.

[10] A. Y. Coran, Rubber Chem. Technol. **1988,** 61, 281.

[11] A. Y. Coran, Rubber Chem. Technol. **1991,** 64, 801.

[12] Y. W. Chang, Y. S. Shin, H. Chun, C. Nah, J. Appl. Polym. Sci. **1999,** 73, 749.

[13] J. Lohmar, Kautsch Gummi Kunstst. **1986,** 39, 1065.

[14] D. K. Setua, J. L. White, Polym. Eng. Sci. **1991,** 31, 1742.

[15] D. K. Setua, J. L. White, Kautsch Gummi Kunstst. **1991,** 44, 137.

[16] D. K. Setua, J. L. White, Kautsch Gummi Kunstst. **1991,** 44, 542.

[17] D. K. Setua, J. L. White, Kautsch Gummi Kunstst. **1991,**44, 137.

[18] D. K. Setua, K. N. Pandey, A. K. Saxena, G. N. Mathur, J. Appl. Polym. Sci. **1999,** 74, 480.

[19] B. Boutevin, E. Fleury, J. P. Parisi, Y. Piétrasnta, Makromo. Chem. **1989,** 190, 2363.

[20] F. Romani, E. Passaglia, M. Aglietto, G. Ruggeri, Macromol. Chem. Phys. **1999,** 200, 524.

[21] U. Gorski, K. Maenz, D. Stadermann, Angew. Makrom. Chem. **1997,** 253, 51.

[22] M. G. Oliveira, B. G. Soares, C. M. F. Santos, M. F. Diniz, R. C. L. Dutra, Macromol Rapid Commun. **1999,** 20, 526.

[23] P. Jansen, M. Amorim, A. S. Gomes, B. G. Soares, J. Appl. Polym. Sci. **1995,** 58, 101.

[24] P. Jansen, A.S. Gomes, B. G. Soares, J. Appl. Polym. Sci. **1996,** 61, 591.

[25] P. Jansen, B. G. Soares, Polym. Degrad. Stab. **1996,** 52, 95.

[26] P. Jansen, B. G. Soares, J. Appl. Polym. Sci. **2001,** 79, 193.

[27] B. G. Soares, F. F. Alves, M. G. Oliveira, A. C. F. Moreira, F. G. Garcia, M. F. S. Lopes, Eur. Polym. J. **2001,** 37, 1577.

[28] V. A. Shershnev, Rubber Chem. Technol. **1892,** 55, 537.

*Macromol. Symp. **189**, 59–81 (2002)*

Morphology Development and Mechanical Response of Supramolecular Structures in Rubber Blends

R. H. Schuster

Deutsches Institut für Kautschuktechnologie e. V.
Eupener Straße 33, 30519 Hannover, Germany
Schuster@DIKautschuk.de

Introduction

The aims of elastomer property design with respect to end-user demands is to arrive at optimum combinations of physical properties common to the originating polymers. For instance, high elasticity, energy dissipation in a given frequency or temperature range and good abrasion resistance are property combinations which have to be achieved in quality tire treads. It has been suggested that energy dissipation in the low frequency region (or at high temperatures) is responsible for rolling resistance of tire treads, whereas energy dissipation in the high frequency region (or at low temperatures) for improving grip properties of treads. Because no polymer alone can fulfil this demand, different rubbers have to be blended. This route offers an economical realization of high performance target properties. The desired property combination does not, however, require miscible blends but some degree of physical autonomy of the phases (heterogeneous phase morphology) [1,2].

In addition the reinforcing effect of particulate colloidal fillers such as carbon black or silica and the success of co-crosslinking the polymer phases determine the technological performance of the final products. Reinforcement by nano-scaled fillers controls especially dynamic and ultimate properties of filled rubber products [3].The blends' physical properties thus do not depend on the size, shape and distribution of the polymer phases alone, but also on filler dispersion, filler transfer and selective distribution in the distinct polymer phases [4]. In heterogeneous blends, the intermolecular interactions demonstrate their efficiency mainly at polymer-polymer and polymer-filler interfaces. Due to these interactions supramolecular structures – like interpenetrating phase networks – are formed in unfilled polymer blends. The aggregation of filler particles, moreover, leads to filler networks that interpenetrate the rubber networks. The self-similar structure of the filler network leads to physical properties which can be described by scaling laws with fractal exponents [5].

The complex behavior arouses considerable scientific and technological interest as it is directly connected to the dynamic and ultimate properties of elastomers. In spite of advances made in relating morphology development during dispersive mixing to rheological factors (i.e. shear rate, viscosity ratio of the matrix and dispersed phase etc.) [6,7], the rather subtle role of interfacial tension has rarely been considered quantitatively in experimental investigations on rubber blends.

          CCC 1022-1360/00/$ 17.50+.50/0

While he development of blend morphology is no doubt directly connected to the processing strategy employed, driving forces of a thermodynamic nature also have to be taken into account. The aim of this contributions to underline the role of the thermodynamic parameters governing the formation of supramolecular structures in rubber blends and in the filler network architecture in relation to their mechanical consequences.

Morphology of unfilled rubber blends

Miscibility of polymers

A fundamental question which should be addressed first with respect to any blend system under consideration is whether the components are miscible or not. While the definition and estimation of compatibility in blends systems is not devoid of ambiguities, the use of thermodynamic criteria is an unambiguous way to judge the case. The necessary conditions for the miscibility of a binary system of a particular composition are connected to the shape of the free energy of mixing as a function of the volume fraction, ϕ:

$$\Delta G_m < 0 \tag{1}$$

$$\frac{\partial^2 \Delta G_m}{\partial \phi^2} < 0 \tag{2}$$

The opposite applies for phase-separated systems.

For blends containing non-polar polymers (i. e. polydienes), the free energy of mixing can be described by applying a formal extension of the Flory - Huggins lattice theory [8,9] proposed by Scott [10]:

$$\Delta G_m = RT \left[\phi_1 \ln \phi_1 + \phi_2 \ln \phi + \chi_{12} \phi_1^2 \phi_2^2 \right] \tag{3}$$

where χ_{12} is the polymer-polymer interaction parameter and x_i the degree of polymerization of each blend constituent.

This expression contains the combinatorial entropy of mixing, which expresses the increase in positional disorder during the mixing process and is a simple function of the volume fractions of each constituent The entropy of mixing always contributes to misciblility. Because no theoretical provision can be made for the enthalpy of mixing, the contribution of this thermodynamic function and the excess entropy of mixing is included in the polymer-polymer interaction parameter χ_{12}, which by definition has the physical meaning of a "residual free energy of mixing." Any knowledge about χ_{12} must be deduced from the experiment. The cohesive energy density concept proposed by Hildebrand offers one possibility for estimating χ_{12} for non-polar systems. It predicts [11]:

$$\chi_{12} = V_r (\delta_1 - \delta_2)^2 / RT \tag{4}$$

where δ_i is the cohesive energy density of the polymer.

The necessary condition for obtaining miscibility is a low-value tendency o the part of χ_{12} or $|\delta_1 - \delta_2|$ ($\chi_{12} \to 0$).The conditions for phase separation obtained equating the binodal and the spinodal conditions lead to the definition of a critical χ_{12}-parameter which is a function of the molar mass. Predictions for miscibility can be made if the critical χ-parameter is considered [12].

In accordance with a lattice concept the interfacial tension γ_{12} is proportionate to the square root of χ_{12} [13].

$$\gamma_{12} \sim \chi_{12}^{1/2} \sim |\delta_1 - \delta_2| \tag{5}$$

This approach makes it possible to predict the interfacial tension and the degree of incompatibility of the blend constituents as well. In a first approximation the thickness of the interface is inversely proportionate to the interfacial tension

$$a_1 =\sim \frac{1}{\chi_{12}} \sim \frac{1}{|\delta_1 - \delta_2|} \tag{6}$$

A useful tool for obtaining χ_{12} and δ for non-polar polymers is the inverse gas chromatography (IGC) [14,15]. The necessary condition for any sucses is the proper selection of the molecular probes and precisely defined experimental conditions [15]. There are certain advantages to thermodynamically selecting well characterized hydrocarbons (i.e. n-alkanes, iso-alkanes, cycloalkanes and aromates) as molecular probes for gas-liquid-chromatography (GLC) on melted polymer films. On the one hand, only dispersive and weak dipole forces contribute to the specific retention values and χ_{12}. On the other hand, the heat of evaporation of these molecular probes has to be precisely known [16]. Therefore δ_2-parameters reflect the enthalpy contributions of this range of interaction forces and entropy effects [15,17].

Influence of molecular weight

It can easily be seen that for various rubber blend combinations the critical molar masses above which phase separation occurs, M_{cr} (calculated for the symmetrical case $M_{cr,2} = M_{cr,3}$), are far below the molecular weights of technical rubber grades. Consequently, irrespective of the mixing technique employed, the result will be heterogeneous phase morphology (Fig. 1). The value of δ_2-parameters as an indication of the degree of non-miscibility was proved for various polydienes and polyolefines [2,18]. The hyperbolic phase evolution diagram (Fig. 1) shows that miscible blends are obtained from technical rubber grades only if the δ-parameter difference is less than $0.2 \text{ J}^{1/2} \text{cm}^{3/2}$

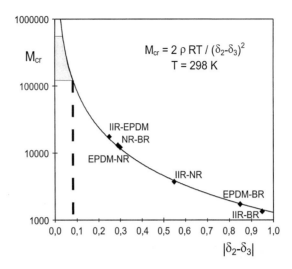

Figure 1: Phase diagram indicating the critical molar mass of the polymer as a function of the δ-parameter difference of rubber blends.

Contributions of microstructure and specific interacting groups

Based on the precise GLC determination of δ-parameters, the thermodynamic contribution to the morphological development of the butadiene microstructure units and phenyl groups was established by covering a large compositional range of BR (cis-BR, vinyl-BR) and SBR (E-SBR; S-SBR) types (Fig. 2).

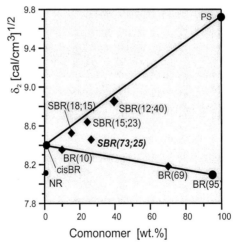

Figure 2:   δ-Parameter of BR-copolymers as a function of the comonomer content.

Although both the vinyl unit and the styrene unit contribute to chain dynamics in a similar way by increasing proportionately the T_g and thus have an influence on damping behavior and grip properties, their thermodynamic behavior is completely opposite. Increasing the vinyl content in the cis-BR decreases polarizability and interaction potential. This leads to a linear decrease in the δ-parameter as vinyl content increases. As a consequence, the degree of compatibility of BR with all polymers with a lower $δ_2$-parameter (i.e. NR, EPDM, IIR, PP, PE) increases. At a vinyl content of 70-80%, the δ-parameter of BR equals that of natural rubber or synthetic polyisoprene [18]. As a result the mixing enthalpy becomes practically zero and the blend demonstrates miscibility as shown by indirect methods [19,20]. This notable example of miscibility of two non-polar homopolymers demonstrates the typical behavior of an athermal mixture ($ΔH_m ≅ 0$; $ΔS_m > 0$). In contrast to the common conclusion drawn for the majority of polar systems in which miscibility is achieved only with high negative mixing heats , this result shows miscibility to be responsible for combinatorial entropy [19]. Because the expansion coefficients are nearly the same for both polymers, the equation of state contribution and excess mixing entropy is negligible. However, the broadness of the glass transition region at equal polymer volume fractions indicates micro-heterogeneity which cannot be predicted by the thermodynamic estimation [2].

A pronounced linear increase in the δ-parameter with increasing styrene content indicates that more exothermic interactions govern the thermodynamic behavior of SBR types. As a result the compatibility with less polar rubbers like BR, NR, EPDM decreases as the styrene content increases. On the basis of these observations it can be stated that randomly distributed chain segments of different chemical constitution deliver specific incremental contributions to the overall thermodynamic behavior of the polymer. As a result, the δ-parameter of any SBR under consideration can be expressed by the parameter equation [3]:

$$δ_{(SBR)} = 17.17 + 0.0272(S) - 0.0069(V) \tag{7}$$

where S and V denote the weight percentage of styrene and vinyl units in the product and δ is expressed in $J^{1/2} cm^{3/2}$.

Equation (7) describes quantitatively the opposite effects of vinyl and styrene units on the thermodynamic behavior of the polymer. However, the effect of blockiness and tacticity are still unknown.

Temperature dependence

As the temperature is varied during the mixing process, the nature of the free energy as a function of composition may change and can result in transitions through the coexistence curve from the solubility gap into the homogeneous region in the phase diagram. For scientific and technical reasons it is interesting to know whether the phase separated system is below a UCST or above an LCST. Because rubbers are blended exclusively by mechanical mixing, the effect of increasing mixing temperature can decisively influence the final phase morphology.

A shortcoming when describing real systems is that there is no theoretical provision made for temperature dependence of χ_{12}. A real advantage of GLC is the estimation of the temperature dependence of δ-
or χ_{12}-parameters. It was experimentally proven that for blends from non-polar rubbers the δ-parameter
difference $|\delta_1 - \delta_2|$, which is directly proportionate to the interfacial tension γ_{12}, is decreasing when the
temperature is increased [21] (Fig. 3). Consequently, by increasing the mixing temperature the interfacial thickness in all non-polar blend systems is increased (Fig. 3).

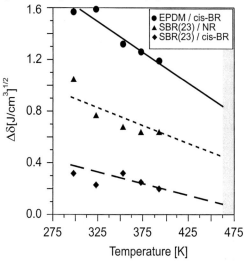

Figure 3:   Temperature dependence of the interfacial tension for non-polar rubber blends.

Despite the fact that producing miscible rubber blends is not a technological target, in most cases the
phase transition into the homogeneous region cannot be achieved due to a residual value of unlike segmental interactions. Partial solubility can, nevertheless, be attained. On the basis of the temperature de-
pendence of $|\delta_1 - \delta_2|$, the existence of a UCST in the temperature region 130-140 ^{0}C is predicted for the
blend SBR(23)/SBR(40). By way of confirmation, the uncured blend system SBR(23)/SBR(40) exhibits
two damping maxima in blending operations at 80 ^{0}C. This indicates a phase-separated system. The
same blend system demonstrates single phase behavior if the mixing temperature passes the critical
value of 130-140 °C[23]. However, reducing the temperature during storage triggers an equilibration
into a two-phase morphology in uncross-linked blends.

Morphology development in heterogeneous rubber blends

Rubber blends with co-continuous phase morphology provide good ultimate properties due to the
three-dimensional interpenetrating phase network formed. Despite the technological advantages of
such systems, blends with one phase dispersed in a continuous matrix aroused the greatest scientific

interest . In the case of such blends, the resulting domain size, d, can be theoretically described as a function of both thermodynamic and the rheological contributions [24]

$$\overline{d} = (c\gamma/\eta_m) \, [dV_x/dy]^{-1} \cdot f \, (\eta_m/\eta_d) \tag{8}$$

where dV_x/dy - shear rate

η_m - viscosity of the matrix phase

η_d - viscosity of the dispersed phase.

Certainly, the development of blend morphology is directly connected to the processing conditions employed and the viscosity of the blend constituents [7,25], but driving forces of a thermodynamic nature also have to be taken into consideration.

Dispersion kinetics

In the early stage of polymer dispersion the phase break-up is related to the stress level exerted on the droplets of the minor component by the flowing matrix [26]. The effectiveness at which stress can sustain particle deformation depends on the degree of interdiffusion of the polymer chains at the phase boundary. Due to the fact that chain length and frictional coefficients of non-polar rubber are not very dissimilar, the interdiffusion, which is governed by the degree of thermodynamic similarity, creates interphases whose thickness could be assumed to be at the same order of magnitude. The phase break-down of the minor component starts with a typical exponential decrease of the domain size and levels at prolonged mixing into a stationary state (Fig. 4).

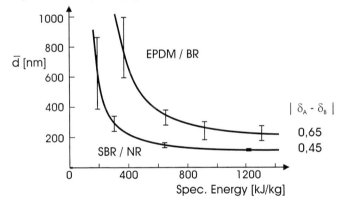

Figure 4:   Domain size reduction during mixing of rubber blends.

If the process is treated as a reaction of the first order, the corresponding reaction rate constants demonstrate a dependency on the thermodynamic dissimilarities of the blend constituents. From a comparison of the dispersion rate of BR in EPDM with that of SBR in NR it is obvious that systems with a higher degree of compatibility tend to form more rapid a stationary phase morphology than those with a poor compatibility.

$$k_{EPDM/BR} = 11.5 \cdot 10^{-3} \text{ [s}^{-1}]$$
$$k_{NR/SBR} = 18.3 \cdot 10^{-3} \text{ [s}^{-1}]$$

These thermodynamically driven dispersion kinetics was confirmed for blends of cis-1,4-BR with SBR containing an increasing amount of styrene units. It was shown that keeping the viscosity ratio η_m/η_d at constant the dispersion rate is a linear function of the δ-parameter difference [21]. More detailed investigations of the domain size reduction by TEM revealed a multimodal size distribution. The value of the ratio of consecutive size maxima is asymptotically lowered from higher values in the early stages of mixing to 1.26 in the stationary mixing stage. This evolution indicates that longitudinal domains (fibrillas) are formed first. These are disrupted into a sequential reducing number of smaller drops, which again form fibrillas in the shear field. The final ratio of 1.26 demonstrates that in this mixing stage the domains are divided into two similar sized parts.

Influence of shear during mixing

As long as the chemical nature and molecular weight of the constituents are kept constant, the shear rate employed during mixing dominates the domain size reduction (Table 1). When the ratio of shear rate to interfacial forces is taken into account, it became obvious from this quantitative data that interfacial interactions are graded similarly to rheological influences.

Table 1: Influence of rotor speed on domain sizes in selected blends.

System	Rotor speed (rpm)	Domain size (nm)	$\|\delta_1 - \delta_2\|$ $(J/cm^3)^{1/2}$
NR/SBR (23)	65	260	0.92
	95	120	
EPDM/BR	65	650	1.33
	95	240	

This basic result for unfilled rubber blends cannot be transferred to filled systems without reservation. When mixing rubbers without fillers, the effects of polymer incompatibility and shear rate are of the same order of magnitude, whereas in the presence of fillers – especially above the percolation threshold – the influence of shear rate is predominant. Domain sizes are also affected by the presence of particulate fillers [27]

Stationary Domain Size

For the most clear-cut picture in which the dispersed phase is surrounded by a continuous one, the relationship between the stationary average domain size and the interfacial tension was investigated for a large number of non-polar rubber blends under properly controlled mixing conditions (constant rotor speed and temperature)[21,27]. For uncured NR/E-SBR blends (7/3) it was observed that the stationary average domain size increases with both the styrene content and the δ-parameter difference (Fig. 5) [28]. Outside of this group the linear dependency of the stationary average domain size on δ-parameter difference was proved for various NR-blends with BR, vinyl-BR, E-SBR, S-SBR and EPDM and for more incompatible blends based on EPDM or CIIR with the polydienes mentioned, in accordance with the equation [18]

$$d = 1260 \, (\Delta\delta - 0.204) \qquad (9)$$

where d denotes the average domain size in nm and δ is given in $J^{1/2} \, cm^{3/2}$.

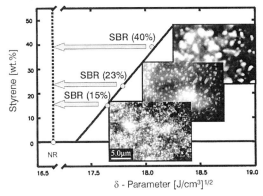

Figure 5:   Domain size vs. δ-parameter difference for various rubber blends.

With increasing shear rate (rotor speed) the slope of d ~ Δδ will decrease. Consequently, the rheological influences will affect blends with a low degree of compatibility to a proportionately higher extent. The temperature in the mix increases because of high shear flow. If the δ-parameter of blend constituents shows a convergent temperature dependency, this effect can offset the decrease of shear stress by bringing about a reduction in melt viscosity. In such a case the decreased interfacial tension will result in a finer dispersion. If the temperature dependencies of δ-parameters are not convergent like they are for NR/E-SBR(23) or NR/SBR(40), the increase in rotor speed, and therefore in mixing temperature, leads to an increase in the average domain size, especially as a result of broader domain size distribution. The effect observed is attributed to a reduction of the matrix viscosity due to both the molecular weight breakdown in the NR phase and the increased temperature [27].

From TEM investigations on NR/SBR blends it became obvious that the aspect ratio (AR) of the domains is increased if the degree of compatibility of the blend constituents is decreased [18]

$$AR = d_{min}/d_{max} = 0.671 + 0.178 \, \Delta\delta \qquad (10)$$

Equ. 10 indicates that in more compatible blends the interfacial interaction is able to sustain domain shapes with a less favorable volume surface fraction than that of a sphere (i.e. ellipsoidal or fibrilla like domains).

68

Phase inversion

Regarding the contribution of interfacial interactions to the concentration region where phase inversion or the co-continuous phase morphology occurs, significant publications have demonstrated that the viscosity ratio influences the location of the phase inversion region [25]. From TEM investigations on NR/SBR blends, preformed in the author's laboratory, it became obvious that the concentration region in which co-continuous phase morphology does appear is enlarged when the degree of compatibility of the blend constituents increases [29] (Fig. 6). However, since the co-continuous structures result in enhancement of mechanical blend properties, more detailed knowledge about phase inversion and its extent can be rewarding.

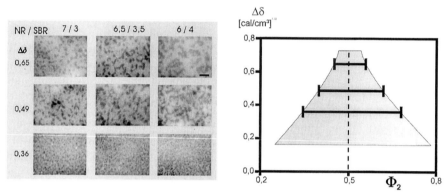

Figure 6:   Concentration range of the phase inversion region as a function of the interfacial tension of the blend constituents (System NR/SBR).

Structure of interpenetrating phase networks

Information about the architecture of co-continuous phase morphologies can be gained from dynamic -mechanical properties of the blend systems in the glass transition region [30]. The dynamic-mechanical response of multi-component rubber blends follows from the general principles of energy conservation, which state that the total energy stored or dissipated in a dynamically excited blend system results additively from the energy contributions of the different phases. Since the energy stored and dissipated per unit volume can be expressed by the storage and loss modulus, G' and G", multiplied by the square of the mean strain, the complex modulus $G^* = G' + i\,G"$ fulfills the relation:

$$G^*_{blend} = \sum_i G^*_{P,i}\,\phi_i \left(\frac{\gamma_{0,i}}{\gamma_0}\right)^2 \tag{11}$$

Here, G^*_{blend} is the modulus of the blend and γ_0 is the external strain, i.e. the mean strain averaged over the whole blend system. Both quantities are macroscopically observable parameters. Since any measurement on a rubber blend averages over the whole sample, evidently. $G^*_{P,i}$ is the complex modulus, ϕ_i the volume fraction and $\gamma_{o,i}$ the average strain of the i-th blend component.

A purely geometrical expression for $\gamma_{o,i}$ can be derived if a critical behavior, e.g. a percolation structure, is assumed for the phase network and the i-th phase is glassy, while all other phases are rubbery. The complex modulus is then governed by the glassy i-th phase and the contributions of the other phases can be neglected, i.e. the sum in Equ. (11) reduces to a single summand. By introducing the glass transition temperature $T_{g,i}$ of the i-th phase, the complex modulus at this particular temperature fulfills a critical scaling law for $\phi_i > \phi_c$:

$$G^*_{blend}(\phi_i; T_{g,i}) = G^*_{P,i}(T_{g,i}) \left(\frac{\phi_i - \phi_c}{1 - \phi_c} \right)^\tau \tag{12}$$

Here, ϕ_c is the critical volume fraction above which a phase network is formed. It equals the percolation threshold if a percolation model is assumed. The elasticity exponent τ depends on the particular structure of the interpenetrating phase network. It is affected mainly by the distribution of flexible connections in the phase network that are not arranged in close loop structures and dead ends.

For G' and G" analogue relations are valid.

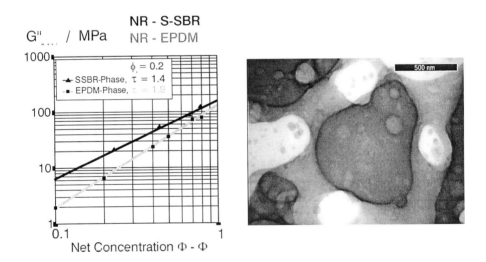

Scaling law G" vs (ϕ_i - ϕ) for NR - S-SBR and NR - EPDM and TEM-micrograph of an IPN (NR - EPDM)

Figure 7: Double logarithmic plot of G" vs. (ϕ_i - ϕ_c).

The double logarithmic plot of G" vs $(\phi_i - \phi_c)/(1 - \phi_c)$ delivers a straight line with the slope τ. The experimental results leads to fractal exponents. Depending on the system the τ-values are in the range of 1.5 to 1.9. The result indicates that the interpenetrating phase network corresponds to a supramolecular organized fractal object which contain a given fraction of stiff loops of dimensions similar to the diameter of a polymer coil. With decreasing degree of compatibility, the exponent also decreases because the diameter of the loops in the blend system becomes smaller (Fig. 7).

Dimension of interphase

The thickness of the interphase and their development during the mixing process and when stored was examined by thermoanalytical methods. By using blends of semi-crystalline cis-BR and a sequence of amorphous polydienes, it was shown that the normalized loss of crystallization increases when the blend constituent demonstrates a higher degree of compatibility [21]. The thickness of the blend interphase a_I was derived from the volume fraction of non-crystallizing cis-BR. The experimental results confirm the theoretically expected relation (6): $\gamma \sim a_I^{-1} \sim |\delta_A - \delta_B|$ (Fig. 8).

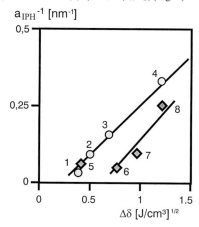

Figure 8: Reciprocal thickness of the interphase vs. interfacial tension uncrosslinked Systems: 1 - cis-1,4-BR/E-SBR (17;23), 2 - cis-1,4-BR/E-SBR (17;29), 3 - cis-1,4-BR/E-SBR (17;40), 4 - cis-1,4-BR/EPDM crosslinked Systems: 5 - NR/cis-1,4-BR, 6 - NR/S-SBR (50;25), 7 - NR/E-SBR (17;25), 8 - NR/EPDM.

With the help of a new method the interphase signal can be obtained from the temperature dependency of the loss modulus maximum [30]. It was shown that the interface of crosslinked blends is larger than of non-crosslinked ones, but that it is also inversely proportional to the interfacial tension. The estimated thickness of the interphase ranges from 1.5 nm for highly incompatible blends (i.e. EPDM/BR) up to 15-20 nm for highly compatible blends (cis-BR/SBR).

Physical properties of unfilled blends

The main motivation for blending rubbers is to improve and enlarge the property spectrum the elastomers exhibit in ultimate use. A deeper knowledge as to the influence of blend morphology on physical proper-ties – especially ultimate ones – is therefore of critical interest. One can observe that both tensile strength and energy at break increases when the average domain size of the dispersed phase decreases during the mixing process (Fig. 9).

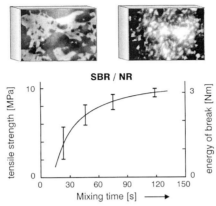

Figure 9:   Evolution of tensile strength and energy at break as a function of mixing time.

Increasing the volume fraction of the interphase by reducing the nominal domain size effects a consider-able improvement in the physical properties. As the volume fraction of the interphase increases, the con-centration of flaws is reduced because of an increased region in which disparate polymer chains are co-crosslinked. The mechanical integrity of the intercrosslinked morphology also leads to greater tear strength than in either of the pure components. This proves that phase morphology of a heterogeneous rubber blend contributes in a determinant way to the physical property level of the final elastomer.

Filled rubbers

The quality of rubber products containing particulate reinforcing fillers is highly dependent on a high degree of dispersion and an optimum distribution of filler particles in the mix [3]. It is, therefore, very important to ensure good dispersion of carbon black or silica in order to obtain products with tailor-made mechanical properties (low hysteresis, high tensile strength, abrasion resistance, cut growth resistance etc.) [31]. Since filler loading is typically high in most of the rubber products (≥ 50 phr) and the mixing time is as short as possible, filler dispersion is often less than optimum. The detri-mental consequences of poor dispersion are [31]:

–   reduced product life,

–   poor performance during service,

- poor product appearance,
- poor processing and manufacturing uniformity,
- waste of raw materials,
- excessive energy usage.

The avoidance of premature product failure was the driving force for the development of better mixing procedures, the investigation of the dispersion influencing factors, a better knowledge of the dispersion mechanism and, last but not least, reliable testing methods for filler dispersion.

The process of dispersion depends: on:

(i)   the chain length of the polymer

(ii)   the share of specifically interacting groups along the chain,

(iii)   the surface activity (interacting potential),

(iv)   distribution of energetically effective adsorption sites

(v)   fractal dimensions of the filler.

Dispersion of fillers is by definition the process of mechanical breakdown of hard filler pellets (commercially available size 1-3 mm) into smaller entities like agglomerates (of 1-50 μm) and, after extended mixing, into aggregates (of 100-300 nm). Distribution means the process in which agglomerates or aggregates are pulled apart and spread out by random patterns of flow in the rubber melt.

A necessary condition for the pellet breakdown is the transmission of shear forces from the mixing device to the filler pellets or agglomerates through the rubber matrix [32]. This can be done only if the rubber matrix exerts a "wetting" influence, or in other words, if there is good interaction (or affinity) between the polymer and the filler [31]. The energetic balance between filler-filler interactions in the agglomerates, on the one hand, and filler-polymer interaction, on the other, is thus controlled by the dispersion process.

The general effects of dispersion on rheological properties are shown in the decrease of Mooney viscosity (Fig. 10)and the evolution of the extrusion shrinkage as a function of mixing time. Viscosity, which is an important processing parameter, decreases markedly as mixing time and rotor speed increase. During the early stage of mixing the voids of the fractal filler aggregates are penetrated by a part of the polymer. After this incorporation and the subsequent dispersion stage the filler aggregates are more uniformly distributed and therefore the melt viscosity is reduced [33].

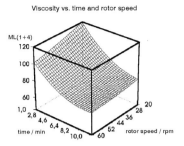

Figure 10:   Decrease of the melt viscosity after mixing as a function of mixing time and rotor speed (ML (1-4) = Mooney viscosity; SBR and carbon black N 326 50 phr [34].

As far as filled crosslinked rubber networks are concerned, the large and poorly dispersed filler agglomerates act as flaws and contribute significantly to failure properties such as tensile strength, elongation, energy at break, fatigue life and abrasion [35]. Properties dependent upon crack initiation and crack propagation are significantly determined by the dispersion and distribution of the filler. As mixing time and rotor speed and therefor the degree of dispersion increase, tensile strength and crack propagation resistance also increase [35](Fig. 11). The loss in tensile fatigue life as a function of decreasing dispersion (or by increasing the frequency of large agglomerates) is common for all elastomer systems.

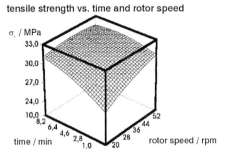

Figure 11:   Evolution of tensile strength of a filled SBR vulcanizate as a function of the mixing conditions [35].

Filler networks

Supra-molecular fractal structures (filler network) are formed as filler concentration ϕ increases. It was demonstrated by using polymeric filler particles ranging in size from 30-400 nm that at low filler concentrations the individual particles first tend to cluster together. Above a critical filler concentration ϕ^* (gel point) the clusters agglomerate and form a three-dimensional filler network by kinetic

74

cluster-cluster aggregation (CCA) [36-38] The filler network obviously interpenetrates the polymer network. Once the filler network is formed the macroscopic stress is no longer transmitted through the rubber matrix alone, but merely through the self similar branches of the filler network [39].The mechanical properties are thus changed in a characteristic manner on all length scales. Because at length scales longer than the size of a cluster, ξ, the filler network can be considered as homogeneous, the modulus of the entire CCA-network equals the modulus of a single cluster. The clusters behave to some degree as elastic units capable to store energy during twisting, bending or stretching the sample. For very small deformation amplitudes above the gel point φ* at which the filler network is built up, the storage modulus G'_o can be expressed [5] as:

$$G'_o \cong G_F \left(\frac{a}{\xi} \right)^{3+d_{f,B}} \cong G_F \phi^{\frac{3+d_{f,B}}{3-d_f}} \cong \phi^{3,5} \tag{13}$$

with d_f =   1.8 - fractal dimension of the filler cluster

   $d_{f,B}$ =   1.3 fractal dimension of a filler cluster

   ξ   =   size of the clusters

   a   =   size of the particles

This scaling law is observed irrespective of the filler type, or the size of the filler particles and the chemical or physical structure of the surface. A typical example is shown in Fig. 12

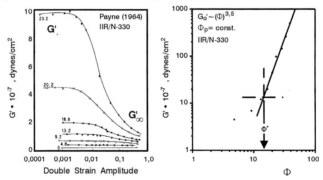

Figure 12:   Influence of the filler concentration on the strain amplitude dependency of G' (low strain).

However, the process of self-organization of the nano-particles is controlled by the interaction of the filler clusters with the surrounding matrix [38]. It was shown that the location of the gel point φ* depends on the intensity of the polymer filler interaction. The smaller this interaction, the greater the tendency to particle cluster formation, so that the gel point φ* is reached at lower filler concentrations. The degree of interaction is basically related to the interfacial tension [40]. A high interfacial tension thus tends to support cluster formation at low filler concentrations. Owing to the scaling law, this again results in a higher storage modulus at constant filler concentration, as in the case of systems with strong polymer-filler-interactions. Cluster formation is avoided in better polymer-filler interaction up to a certain degree. This contributes to better dispersion and mixing quality, which is directly evident in the dynamic and ultimate properties.

However, the filler network is not too stable. Deformations of filled elastomer bodys lead to a decrease of the Van der Waals forces which hold the cluster aggregates together. By increasing the strain amplitude in a dynamic test the values of G' are decreasing in a characteristic manner. This non-linear viscoelastic behaviour is typical for filled elastomers and well known as "Payne effect" [41].

The breakdown of the filler network can be modelized by treating it like a percolation lattice [42]. By increasing the deformation strain a random fracture of single clusters and disconnection of clusters in the space -filling configuration of the CCA-network can be assumed. Therefore the mechanical stress can be transmitted only through the remaining clusters of the network. Modeling the strain dependency of the number of surviving clusters the strain dependent modulus can be approximated by a function of the Havriliak-Negami type[42]

$$G'_{\varepsilon \to 0} = (G'_0 - G'_\infty)[1+(\varepsilon/\varepsilon_{1/2})^\beta]^{-\tau} + G'_\infty \tag{14}$$

where $\tau = 3.6$ is the elasticity exponent of percolation,β is an empirical exponent and $\varepsilon_{1/2}$ is the strain amplitude at which half of the clusters in the lattice are vanished at high deformations.

The drop of G' with increasing strain amplitude corresponds to a gel - sol transition of the filler network at moderate strain and a breakdown of remaining sub-clusters at larger strain. The first process can be described by a monotonic function which reaches the level of hydrodynamic effects of dead ends and broken sub-clusters. The second contribution passes through a maximum because the resulting sub-clusters exert furthermore hydrodynamic reinforcing effects. This contribution become larger and passes through a maximum when the sub-clusters are destroyed. The summation of both contributions leads to the shape of the strain amplitude dependency shown in Fig. 12.The breakdown of the filler network as the dynamic deformation amplitude increases is paralleled by energy dissipation [43]. The process leads to a maximum value of G" at the strain amplitude where G' shows a characteristic inflection point. The position of the G" maximum and the absolute value of G"$_{max}$ depend on the balance of the filler-filler and the filler -matrix -interactions. If one of these influencing parameters are controlled by chemical or physical procedures energy dissipation can be triggered on with useful consequences in application.If the strain amplitude dependence of G' and G" are considered for different types of fillers the influence of the filler surface activity become obvious. Fillers with efficient interparticle and interaggregate interactions demonstrate a high tendency to agglomerate and show the most pronounced Payne effect. If the polymer segments are able to respond with similarly graded interactions, the balance of inter-aggregate and filler-matrix interactions can lead to a higher degree of dispersion, a lower value of the gel point $\phi^*$and of G'$_0$. Consequently, any change in surface

76

activity leading to reduced filler-filler interaction result in less pronounced Payne effect. Therefore, the major factors to minimise the difference $G'_0(\omega)$-$G'\infty(\omega)$ are:

(i)   suitable surface activity

(ii)   better polymer-filler interaction

(iii)   high structure of the filler

(iv)   effective mixing

Filler surface activity

Phase bonding between polymer and filler particles is frequently realized by adsorption of polymer segments on the filler surface and deliberately established via functional groups or bifunctional cross-linkers by chemical bonds. Recently it was demonstrated that on the surface of carbon black particles four types of adsorption sites with discrete energy levels occur. The amount of high energetic sites decreases with the increasing primary particle size[44] Because the high energetic sites on the filler surface are supporting stronger interaggregate interactions it become clear why carbon blacks with high surface specific area show poor degree of dispersion and large Payne effect [3].

The surface activity of carbon blacks can be changed by controlled surface treatment which reduces the concentration of the high energy sites. This weakens the interaggregate interactions and promotes filler-matrix interactions. As can be seen from the Cole-Cole plot in Fig. 13, this leads to a minimization of G' and especially of G''. Therefore interaggregate interactions which give rise to a high portion of energy dissipation with amplitude dependent dynamic deformation are effectively reduced.

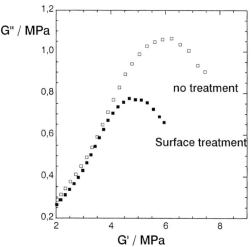

Figure 13: Cole-Cole plots of SBR vulcanizates filled with untreated and treated carbon black (each point corresponds to a given strain amplitude starting from the right with the lowest ε values).

Compared to carbon black, silica can form stronger and more developed agglomerates which would be reflected in higher storage modulus at low strain amplitude and larger Payne effect at low and high temperatures. In a hydrocarbon polymer the polymer-silica interaction which is determined by the dispersive component of filler surface energy is lower than the inter-particle interaction which is supported by the polar silanol groups. Consequently the poor compatibility with the polymer and the strong interaction filler – filler interaction due to the high surface activity lead to a more developed filler network. Changing the surface activity of silica by treatment with mono- or bifunctional silanes the Payne effect can be gradually reduced to a desired level (Fig.14)

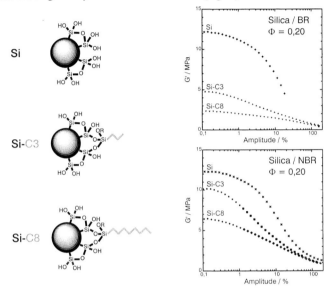

Figure 14:   Payne effect for filled BR-compounds with surface treated silica.

Such changes in surface energy distribution lead not only to a better filler dispersion and lower G' and G" but also to a more pronounced energy dissipation in the Tg region. In other words, the better the filler-matrix interaction the higher the amount of energy necessary to convey the same degree of mobility to polymer chains attached to the filler surface or entangled with such chains. Keeping in mind that energy dissipation in the T_g region is a high frequency phenomenon connected with grip properties the possibilities for monitoring complex physical processes by adjusting the surface interaction potential become clear

In conclusion, changing the surface energy distribution of the filler during the production process or by physical or chemical post treatment has a profitable effect on the filler network formation, the percolation threshold and the placement of the energy dissipation maximum. Therefore energy dissipation can be influenced in the low temperature region mainly by selecting the proper rubber matrix and in the high temperature region by adjusting the surface energy distribution of the filler surface.

The intensity of rubber-filler-interactions can be influenced also by icorporating specific monomers and selective interacting groups in polymer chains. As competing absorption experiments have demonstrated, segmental interaction towards carbon black increases in the following order [18]:

Propylene < Ethylene < 1,4-Isoprene $\cong$ 1,2-Butadiene < 1,4-Butadiene << Styrene

With increasing the content of better interacting monomer units in a polymer chain the percolation limit is shifted towards larger filler concentration, the degree of dispersion is increased and the difference $G'_0 - G'_\infty$ as well as G'' is reduced. By introducing functional groups able to specifically interacting with the filler the energy dissipation in the low temperature region is significantly increased [45].

Filler distribution in rubber blends

One major consequence of the polymer - filler interaction is the particular distribution of the filler in rubber blends.With decreasing polymer viscosity, or molar mass, the influence of segmental interaction-capacity with the active centers of the filler on the preferential distribution of the filler increases. Based on a recently developed dynamical-mechanical method, the filler distribution is determined by recording the influence of the filler on the energy dissipation in the Tg-region [43]. Quantitative results show that the filler transfer occurs during mixing and the distribution of the filler in polymer blends is equivalent to an equilibrium. It was demonstrated that filler partition occurs between the two discrete phases and the interphase. Except for cases in which the filler-matrix interaction is preferentially oriented towards one polymer, a reinforcement of the interface and, at the same time, an increase in their volume fraction is observed (Fig. 15).

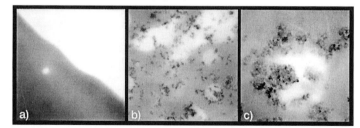

Figure 15:   Distribution of surface modificated silica in EPDM/NBR blends. a) unfilled blend b) EPDM/NBR/silicanized silica c) enlarged view of an EPDM domain surrounded by surface modificated silica.

The phase morphology can be adjusted selectively either by functionalizing filler and polymer as well as by changing the mixing order. Especially interesting are those cases in which interpenetrating polymer networks are reinforced by the filler network that is distributed in the phase interface. By recording the dynamic mechanic response for filled NR/S-SBR in the Tg-region it was shown that the majorpart of the filler is located in the interphase (Fig. 16). As a consequence the volume fraction of the interphase and the interphase signal (between the two damping maxima) become significanlty larger. Especially such effects can be exploited for designing high frequency energy dissipation and grip properties of tire treads [43].

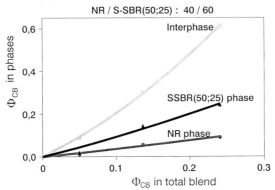

Figure 16:   Concentration of CB in the blend phases and the interphase.

Mechanical consequences from filler distribution

Carbon black distribution exerts a significant effect on dynamic modulus at low strain amplitudes (non linear dependence) . The low-strain modulus increases with an uneven distribution of carbon black between the blend components. As for energy dissipation, it was shown that in the case of NR/SBR-lends, a higher loading of the NR-phase reduces heat build-up and increases rolling resistance It was concluded that hysteresis can be minimized by:

(i)   large carbon black agglomerate size and broad size distribution,

(ii)   higher carbon black loading in discrete domains,

(iii)   selecting the polymer with lower hysteresis as the continuous phase.

In blends of NR with either SBR or BR, lower tensile strength has been reported with a lower carbon black loading in SBR or BR. The most pronounced loss of tensile strength was observed when all of the carbon black was added to the NR-phase [31]. For SBR/BR-blends, a lower-tensile pattern has been observed with higher carbon black in the SBR-phase. Usually the changes caused by phase distributions are found to be greater than the differences between carbon black grades.

The main influences on tear resistance came from

−   small carbon black aggregate sizes

−   low filler structure

−   higher loading of the continuous phase,

−   the polymer with higher strength than the continuous phase.

Conclusions

The configuration of supramolecular structures in phase-separated rubber blends is determined by the type and frequency of intermolecular interaction. The degree of compatibility and the interfacial tension can be reliably described with the help of the solubility parameter determined by means of inverse gas chromatography. Without neglecting the influence of rheological factors, it can be shown that the concentration range of the phase inversion region is determined by the interfacial tension. The weaker the compatibility of the components, the narrower the concentration range in which co-continuous phase morphologies occur. The co-continuous phase morphologies correspond to supramolecularly organized fractal interpenetrating phase networks, the structures of which influence the dynamic-mechanical properties in the glass-transition region. Phase bonding, which depends largely on the – likewise predictable – thickness of the interphase, decisively influences ultimate properties.

Analogous thermodynamically motivated mechanisms determine the rubber-filler interaction. Under identical mixing conditions, the dispersion of the fillers depends on the intensity of the adsorptive interaction between polymer segments and active sites of the filler. The degree of dispersion attained has a crucial influence on the magnitude of energy dissipation at low and high frequencies, on the one hand, and on the ultimate properties, on the other hand. Filler distribution is, moreover, a direct consequence of the respective interaction between the filler and the chain segment of the polymers. The dispersion obtained after long mixing corresponds to the states of quasi equilibrium obtained and has a crucial influence on the mechanical behavior of filled elastomers.

References

[1]   L. Bohn, Angew. Makromol. Chem. **28/30** (1973) 25
[2]   R. H. Schuster, Angew. Makromol. Chem. **202/203** (1992) 159
[3]   G. Kraus, "Reinforcement of Elastomers", Intersc. Publ., New York (1965)
[4]   W. M. Hess, C. R. Herd and P. C. Vegvari, Rubber Chem. Technol. **66** (1993) 329
[5]   M. Klüppel, R. H. Schuster and G. Heinrich, Rubber Chem. Technol. **70** (1997) 243
[6]   L. Utracki, "Polymer Alloys", New York, Hanser Verlag (1999)
[7]   N. Tokita, Rubber Chem. Technol. **50** (1977) 292
[8]   P. J. Flory, J. Chem. Phys. **9** (1941) 660
[9]   M. L. Huggins, J. Chem. Phys. **9** (1941) 440
[10] R. L. Scott, J. Chem. Phys. **17** (1949) 268
[11] J. H. Hildebrand and R. L. Scott, "The solubility of non-electrolytes", Dover, New York (1964)
[12] O. Olabisi, L. M. Robeson and M. T. Shaw "Polymer-polymer Miscibilty" Acad. Press, Inc., San Diego, New York (1979) Ch.2
[13] E. Helfand, Chem. Phys. **69** (1975) 2192
[14] J. R. Conder, C. L. Young "Physicochemical Measurements by Gaschromatography (1982)
[15] R. H. Schuster, H. Gräter and H. J. Cantow, Macromolecules **17** (1984) 619
[16] B. D. Smith and R. S. Rivastava, "Physical Science Data, Thermodynamic Data for Pure Compounds, Part A," Elsevier, New york ,1986.
[17] O. Smidsrod and J. E. Guillet, Macromolecules **2** (1969) 279
[18] R. Schuster, H. M. Issel, and V. Peterseim, Rubber Chem. Technol. **69** (1996) 769

[19]   C. M. Roland, Rubber Chem. Technol. **62** (1989) 456

[20]   J. Weik, PhD Thesis Universität Hannover (1993)

[21]   R. H. Schuster, J. Schaper, and M. L. Hallensleben, 155th Meeting of the ACS-Rubber Division, Montreal, April 1997

[22]   M. R. Kamal, R. Lai Fook and N. R. Demarquette J. Polym. Eng. Sci., vol 34(1994) 1834

[23]   R. H. Schuster, 2nd Seminar on Elastomers, Akron (OH), (1993)

[24]   W. Berger, H. W. Kammer, C. Kummerlöwe, Makromol. Chem. Suppl. **8** (1984) 101

[25]   G. N. Avgeropoulos, F. C. Weissert, P. H. Biddison and G. G. A. Böhm, Rubber Chem. Technol. **49** (1976) 93

[26]   Van Oene in " Polymer Blends" (D. R. Paul and S. Newman, Ed.) Vol. 2 Ch7

[27]   V. Peterseim, PhD Thesis, Universität Hannover (1995)

[28]   H.-M. Issel, PhD Thesis, Universität Hannover(1993)

[29]   R. H. Schuster, Worldwide Amazon Conference, Manaus Nov. 1999

[30]   R. H. Schuster, J. Meier and M. Klüppel, Kautsch. Gummi Kunstst. **53** (2000) 663

[31]   W. M. Hess, Rubber Chem. Technol. **64** (1991) 386

[32]   I. Manas-Sloczover, A. Nir and Z. Tadmor, Rubber Chem. Technol., **55** (1982) 1250

[33]   A. I. Medalia, Rubber Chem. Technol. **59** (1986)432

[34]   R. H. Schuster, H. Geisler and R. Staeber , IKV-Seminar "Innovative Technologies in Elastomerprocessing" Aachen, Oct. (1996)

[35]   A. N. Gent, in "Science and Technology of Rubber" (J. E. Mark, B. Erman and F. R. Eirich, Eds.) 2nd Edit. Academic Press (1994) Ch. 10

[36]   A. Bischoff, PhD Thesis Universität Hannover (1992)

[37]   A. Bischoff, M. Klüppel and R. H. Schuster, Polymer Bulletin **40**(1998)283

[38]   R. H. Schuster and M. Müller, German Rubber Conference dkt' 2000, Nürnberg

[39]   P. Meakin, Prop. Solid State Chem. **20** (1990) 153

[40]   R. H. Schuster and M. Müller, 159th Meeting of the ACS-Rubber Division Meeting, Providence (RI), April 2001

[41]   A.R. Payne, J. Polym. Sci.,**6** (1962) 57

[42]   R. H. Schuster, J. Schramm and M. Klüppel, 153th Meeting of the ACS-Rubber Division Meeting, Indianapolis(IN) April 1989

[43]   J. Meier, M. Klüppel and R. H. Schuster, [th] Meeting of the ACS-Rubber Division, Dallas (TX) April 2000

[44]   W. M. Hess and V. E. Chirico, Rubber Chem. Technol. **50** (1977) 301

[45]   R. H. Schuster, ACS-Natl. Meeting San Francisco (CA) March 2000

Macromol. Symp. **189**, 83–103 (2002)

Conductive Polymer Blends: Preparation, Properties and Applications

Marco-A. De Paoli[1]* and Wilson A. Gazotti[2]

[1] Laboratório de Polímeros Condutores e Reciclagem, Instituto de Química, Universidade Estadual de Campinas, C. Postal 6154, 13083-970 Campinas-SP, Brazil
[2] Laboratório de Ciências Químicas, Universidade Estadual do Norte Fluminense, Av. Alberto Lamego 2000, 28015-620 Campos dos Goytacazes-RJ, Brazil

Summary: This article presents a brief review concerning the production of conductive polymer blends and composites which combine conducting and insulating polymers. Different strategies for preparing these mixtures are examined, with emphasis on several properties of the final mixtures, such as: electronic conductivity, mechanical behaviour and thermal stability. The advantages of using blends in some technological applications instead of pure conducting polymers are discussed.

Historical Approach

After Shirakawa's pioneering work[1] showing that polymeric materials could exhibit reasonable electronic conductivity, the research in this field has developed intensely. Nowadays, organic polymers with intrinsic conductivity are a well-known and studied class of materials. Indeed, some conducting polymers are now being produced on a commercial scale by several industries.[2-5] Besides their intrinsic electrical conductivity, they also present electroactivity, electrochromism and semiconductor behaviour, which qualify them for several applications, such as: batteries[6], capacitors[7], electrochromic displays[8-10], field effect transistors [11,12], light emitting diodes[13], photovoltaic [14] and photoelectrochemical cells.[15-17]

Despite these interesting properties, the difficulty of processing conducting polymers by the usual methods of the polymer industry (extrusion, injection moulding, lamination, etc.) restricts their large scale use. Thus, several studies have focused on the synthesis of thermally processable conducting polymers.[18-21] Sulphonic acids, such as dodecylbenzenesulphonic and camphorsulphonic acids, are largely used to obtain polyaniline, PAni, with suitable mechanical properties.[22-25] However, the most

 CCC 1022-1360/00/$ 17.50+.50/0

frequently used method to produce conductive materials with this property is by mixing conducting polymers with conventional processable polymers.[26]

The technology of producing polymer mixtures has become a very important route for obtaining new materials, by combining well-known polymers. Indeed, in recent decades this method has been more frequently used by the polymer industry than research and development on new polymeric materials.[27] The aim of making these mixtures, historically called *polymer blends*, is the synergistic combination of the properties of their components. With mixtures of electronically conductive and insulating polymers, the main objective is the preparation of polymeric materials with good rheological properties and processability associated with electronic conductivity, to produce conductive films for static charge dissipation or electromagnetic interference shielding applications.[28] Other properties inherent to conducting polymers, however, could be affected by the insulating matrix, such as electrochromism and redox behaviour. The field of producing conductive blends by mixing conducting and insulating polymers has been widely studied and several reviews discuss this subject.[29-31] The most commonly used polymers are polypyrrole, polyanilines and polythiophenes, because of their easy preparation and environmental stability.

Mixtures of insulating and conductive polymers are called *polymer blends* or *composites*, however, the latter name is more often used when the insulating polymer is basically a host matrix to the template synthesis of a conducting polymer. For this reason, in this article, the term *conductive blend* is used to designate all types of conductive polymer mixtures, independent of the miscibility between the components or the preparation method.

Conductive polymer blends were first prepared by the electrochemical polymerisation of pyrrole in a poly(vinyl chloride), PVC, modified platinum electrode.[32,33] A PVC film adherring to a platinum plate was swollen by the electrolyte solution containing the pyrrole monomer and, when a suitable potential was applied between this electrode and a reference electrode, anodic polymerisation started from the polymer | electrode interface and propagated to the polymer | electrolyte interface. The association of the electronic properties of the conducting phase with the mechanical properties of the insulating phase produced a new material with thermoplastic-like mechanical properties and electrical conductivity in the range of 10 to 10^2 S cm^{-1}.

In another pioneering work, blends of polypyrrole, PPy, and PVC or brominated polyvinylcarbazole were prepared by potentiostatic polymerisation of pyrrole into the insulating films.[34] The product showed conductivity of 2 S cm^{-1} and 60 % transmittance at 633 nm. According to the authors, the bleached reduced blend was more stable than pure polypyrrole prepared under the same conditions. Nowadays, several other methods are used to prepare conductive blends. These have a significant effect on the final properties of the mixtures. In the follow section, a brief discussion of preparation methods is presented.

Preparation Methods

Electrochemical synthesis: the electrode coating method

As mentioned above, the electrochemical synthesis of a conducting polymer into an insulating film was the first method used to prepare a conductive polymer blend.[32,33] Basically, the working electrode is coated with an insulating polymer, which can be swollen by the electrolytic solvent, and, by the application of a suitable potential or current, the monomer polymerises into the matrix. For this preparation, a conventional electrochemical apparatus (potentiostat/galvanostat, three electrodes/one compartment electrochemical cell, electrodes and an electrolyte containing the monomer) is required. Since the working electrode is previously coated with the insulating film, this method is called the *Electrode coating method.*

Depending on the miscibility, the conducting polymer can be formed within the insulating matrix or between the electrode and the insulating film. The first result occurs if three conditions are fulfilled by the insulating matrix: (i) swelling by the solution containing the monomer and the electrolyte, (ii) a certain degree of miscibility with the conducting polymer and (iii) stability in the potential range used and in the electrolytic medium.

Initially, the electrode coating method was the most used alternative to produce conductive polymer blends.[35-39] Despite the possibility of large scale production of conductive films using a rotating-drum electrode [40], electrochemical methods are usually restricted to small-scale preparations and are presently used only in work related to technological applications where the conductive film must necessarily be deposited on conductive substrates.[41] Thus, the most recent work in this area describes preparation methods which enable large-scale production.

Chemical methods

In situ chemical polymerisation of the conducting polymer into an insulating film is very similar to electrochemical methods, but the potential needed to promote the formation of the conducting polymer is supplied by an oxidising agent. In comparison to electrochemical methods, the main advantage of this method is that the dimension of the conductive film is not limited by the electrode area.

Polymerisation can be achieved by adding the monomer to a solution containing the insulating polymer and an oxidising agent. Blend films are obtained by subsequent evaporation of the solvent. Nicho and Hu used this method to produce blends of PPy with poly(vinyl alcohol) and poly(vinyl acetate). These authors dispersed pyrrole in a solution containing the insulating matrix and ferric chloride. After solvent evaporation, the resulting films showed electronic conductivity of *ca.* 10^{-3} S cm^{-1} and good miscibility between the conductor and the insulating components.[42]

The final evaporation step can be eliminated if the oxidant is incorporated into an insulating film, with subsequent exposition of this film to vapours of the monomer. The solid inorganic oxidant can be introduced into the insulating film in several ways, including mechanical mixing and co-dissolution. Cupric chloride was first incorporated into ethylene-propylene-ethylidene norbonene terpolymer, EPDM, by calendering and the filled insulating films were exposed to pyrrole to obtain an EPDM-PPy blend.[43] Previous incorporation of the oxidant by co-dissolution was used by Mano and co-workers for preparing films by evaporating a tetrahydrofuran solution of PVC and ferric chloride with subsequent exposition to pyrrole vapours.[44]

Chemical polymerisation of the conducting polymer into the polymer matrix can also be done by including the monomer into the polymer matrix and subsequent exposition of the resulting film to an oxidising agent. Green-transparent and conductive mixtures of PAni/poly(methyl methacrylate), PMMA, and PAni/PVC were obtained by dissolving aniline and the host matrix in a common solvent and, after evaporation, immersing the resulting film in an aqueous ferric chloride solution.[45] In another work, Lee and Hong coated porous Nylon membranes by immersing them in a pyrrole solution, with the subsequent addition of a ferric chloride solution to the reaction medium.[46]

In chemical methods, one of the reactants, the monomer or the oxidant, is first added to the insulating host matrix. Thus, the polymerisation kinetics for the conducting polymer depend on the diffusion coefficient of the second reactant into the insulating host. Also, formation of the conducting polymer near the surface of the insulating host may preclude further diffusion of the reactant into its bulk, leading to more heterogeneous systems.

Mechanical methods

Research in the field of conducting polymers has attracted great interest because of the several technological applications proposed for these materials. However, the conducting polymers initially developed were highly infusible and insoluble, restricting their use on a large scale. Great effort was devoted to solve this problem and the statement that polyaniline could be melt-processable and soluble in some solvents by using organic acids as dopants[22-24] increased the possibilities of producing conductive polymer blends by different methods. Soluble polyanilines may also be obtained by polymerising *ortho* substituted anilines.[47-49] For these reasons, polyaniline and its derivatives are the conductive polymers most frequently used to prepare conductive blends by the mechanical mixture of the blend components.

Blend components can be mixed by two methods: (i) using the common machinery of the plastics industry, such as a double screw counter-rotating mixer or a double-screw extruder, and (ii) by co-dissolution. The conducting polymer used in these methods is usually obtained by chemical synthesis, because a larger amount of material can be produced by this route.

Co-dissolution means dissolving the polymeric components in a common solvent and evaporating or spinning the resulting mixture over a suitable substrate to obtain a film. In general, undoped (deprotonated) polyanilines are more soluble than the parent protonated polyaniline, and an alternative to produce conductive blends is to dissolve the emeraldine base and the polymer host in a solvent. After solvent evaporation, the blend film is exposed to vapours of an acid dopant. Attempts have also been carried out to obtain polymer blends from doped polyanilines.

A systematic study about the solubility of polyaniline and some of its alkoxy-substituted derivatives showed that the acid dopant has a strong influence on the solubility of these polymers in several solvents.[50] In this sense, polyanilines doped with camphor sulphonic acid, PAni-CSA, and with dodecylbenzenesulphonic acid, PAnis-DBSA, are soluble in different solvents and blends of this polymer with several thermoplastics and elastomers have been prepared by co-dissolution in *m*-cresol[51] or chloroform.[51-54]

Poly(o-methoxyaniline) doped with different acids is highly soluble in dimethylformamide, DMF, and blends of this polymer with polyurethane[55], polyacrylonitrile[56] and poly(ethylene oxide-co-epichlorohydrin)[57] were prepared by co-dissolution of the polymers in DMF.

The co-dissolution method can also be used to obtain conductive fibres. Norris and co-workers dissolved poly(ethylene oxide) and PAni-CSA in chloroform and ultrafine conductive fibres, with diameters less than 2 μm, were produced by electrostatic spinning.[53]

In a recent work, Ahmed and co-workers[58] showed that specific substances may act as both solvent and dopant for the conducting polymer. These authors dissolved undoped poly(o-toluidine)/poly(m-toluidine) and PMMA in formic acid and obtained a thermoplastic blend containing the conducting polymer in its protonated form. The same strategy was also used with aqueous methanesulphonic acid (70 % by wt) to produce a blend of PAni and poly(4-vinyl pyridine).

As stated above, co-dissolution is one of the most widely used methods to prepare conductive films based on polyanilines. By this method it is possible to produce films with larger dimensions than using in situ electrochemical or chemical methods. However, solvent evaporation of large-areas films is not a suitable method to produce conductive blends on a commercial scale. For these applications, mechanical mixing of the polymers by the conventional machinery present in the polymer industries is preferred.

From the industrial point of view, a thermally processable conducting polymer must present the following properties: (i) appropriate rheological parameters at the temperatures typically used for the host insulating matrix; and (ii) no changes in its electronic conductivity with thermal degradation.[20]

Thermoplastic conductive polymer blends of PAni and a series of insulating hosts (PVC[3,20,59], polyethylene[3], polypropylene[3], polystyrene[3], Noryl™[60] and poly(ethylene terephtalate)[60]) have been produced by different thermal processing methods, ranging from miniature extruders[3], Brabender mixers[59], Rheometers equipped a counter-rotating mixer[60] and hot pressure[61], to pilot-plant scale production.[3]

Elastomeric conductive blends have also been produced by mechanical processing. Vallim and co-workers mixed PAni and nitrilic rubber by calendering and produced elastomeric films by vulcanising the rubber under heating and pressure.[62] EPDM and PAni were mixed in an internal counter-rotating mixer coupled to a Torque Rheometer and the resulting mixture was vulcanised under heat and pressure to obtain conductive films.[63,64]

Mixing temperature, design of the screw and rotating speed are some variables which must be adjusted to blend polymers. Temperature should permit a good flow of the molten polymers, but remain low enough to prevent thermal degradation of the components. The design of the screw should provide the shortest possible residence time. Rotation of mixer blades or the extruder screw must be adjusted to induce good mixing and avoid degradation by shearing.

Mechanical mixing of the blend components is the best alternative to produce conductive blend films on a large-scale. The recent progress achieved in obtaining melt-processable conducting polymers, which can be extruded, mixed or pressed by ordinary machinery present in the polymer industries, makes possible their real utilisation as raw-material for producing conductive flexible films for several technological applications.

Other methods

Lack of solubility is not a problem in preparing conductive blends: an adaptation of the co-dissolution method was recently proposed by Haba and co-workers, who prepared blends of PAni doped with dodecylbenzenesulphonic acid, DBSA, by mixing aqueous dispersions of PAni-DBSA and several insulating hosts, with the subsequent evaporation of water.[65] In a similar way, Gospodinova and co-workers prepared a highly conductive and transparent polymer blend by dispersing very fine colloidal particles of PAni in a poly(vinyl alcohol) solution.[66]

As discussed above, several routes exist to prepare conductive polymer blends. The choice of using one methodology instead of another must take into account the final properties of the resulting mixtures. Conducting polymers and insulating matrixes are usually imiscible or partially miscible. Complete miscibility is undesirable because the conductive network needed to promote the conductivity is not formed if this situation occurs. For the same components, different methods lead to mixtures with. different degrees of compatibilization. However, all methods lead to the formation of mixtures with the necessary compatibilization to produce free-standing films.

Also, the preparation method should be chosen by taking into account the applications where the blend will be used. For example, *in situ* electrochemical polymerisation is usually preferred in technological applications where a modified electrode is required, mechanical methods are preferred if large-scale production of conductive films are needed.

Properties: conductivity, mechanical behaviour and thermal stability

Electronic conductivity

The first objective of adding a conducting polymer to an insulating host is to increase the electronic conductivity of the host. For the production of conductive films, the conducting polymer content needs to reach the percolation threshold to cause the onset of electronic conductivity. This concentration limit is markedly lower than that observed with inorganic-filled insulating polymers. According to theoretical studies, this situation occurs when the conductive filler represents 16 % (v/v) in the mixture [67]. Several examples of conductive polymer blends show the low percolation threshold achieved by using conducting polymers.[68-70] Flexible conductive films prepared by casting a DMF solution containing poly(o-methoxyaniline) doped with p-toluenesulphonic acid, PoAnis-TSA, and poly(ethylene oxide-co-epichlorohydrin) presented the onset of the conductivity with ca. 2 % (v/v) of the conducting polymer.[57]

Yoon and co-workers[71] demonstrated that, for the system PMMA/PAni, it is possible to produce a fibrilar network of the conducting polymer when adding 0.5 % (v/v) of polyaniline to the insulating matrix. The authors interpreted the low percolation threshold in terms of the formation of an interpenetrating network between the components of the mixture.

Conductive polymer blends with ultra-low percolation thresholds have also been produced. Juvin and co-workers described the preparation of a blend by mixing PAni-CSA, PMMA and a plasticiser, obtaining a percolation threshold of 0.041 wt%.[72] In another work, Fraisse and co-workers produced a conductive polymer blend where the addition of 0.03 % by weight of conducting polymer to the insulating matrix promoted the onset of electrical conductivity.[73]

Recent advances in this area are leading to the preparation of highly conductive films, mainly based on polyaniline blends. Yang and Lee added only 5 wt % of polyaniline to a water soluble polyurethane and observed an electrical conductivity of 5 S cm^{-1}.[74] Barra and co-workers[75] obtained conductivity values of 1 S cm^{-1} in blends of polyaniline doped with dodecylbenzenesulphonic acid, PAni-DBSA and poly(ethylene-

co-vinyl acetate), EVA. According to the authors, the protonation degree of the conducting polymer is more effective in the mixture than in pure polyaniline. In another recent work, Makela and co-workers achieved 10 S cm^{-1} with a blend of PMMA and PAni-CSA, containing 40 wt % of the conducting polymer.[76]

Hopkins and Reynolds[77] observed that, for mixtures of polyaniline and amorphous or crystalline Nylon-6, the electrical conductivity is 10 times higher in blends using crystalline polyamide (1 S cm^{-1}, for the blend containing 10 wt% of polyaniline). The authors associated these interesting results to the formation of more organised conductive pathways in the blend containing the crystalline host. This important feature opens the possibility of producing films with very high conductivity.

The electronic conductivity observed in a polymer blend is usually lower than that observed for the pure conducting polymer used to product this blend. However, Han and Im described, in an interesting work, that sometimes it is possible to increase the conductivity by producing a polymer blend.[78] These authors observed that the electronic conductivity of a polymer blend composed of PAni-CSA and a polyimide is higher than that observed for the pure polyaniline doped with the same acid. According to the authors, this effect results from the annealing of polyaniline chains, induced by the insulating matrix.

Mechanical properties

As mentioned, the aim of making conductive polymer blends is to combine the electronic properties of the conducting polymers with the mechanical properties of the insulating polymers. Compared to the pure conducting polymers, blends always present enhanced mechanical properties. However, compared to thermoplastics or elastomers, conductive polymer blends often show loss of thermoplastic/elastomeric behaviour. This undesirable effect is usually proportional to the conducting polymer content in the mixture.

Depending on the conducting phase content, the mechanical properties of the insulating host can be retained. Mixtures of poly(ethylene oxide-*co*-epichlorohydrin) and PoAnis-TSA containing 10 wt% of conducting polymer showed Young moduli and elongations at break similar to those observed for the pure host.[57] This behaviour was also observed in blends of EPDM and polyaniline containing the same concentration of conductive polymer.[63] An interesting phenomenon observed in elastomeric conductive blends is the reinforcing effect induced by the conductive phase in the elastomer.[62] This effect was observed for EPDM-PAni blends and, after hot pressing without a vulcanising agent, the presence of the conducting polymer inhibited elastomer solubilisation and produced a cross-linked rubber.[64] On the other hand, Faez and De Paoli, in a recent work, showed that processing PAni-DBSA in an internal mixer with a small concentration of EPDM rubber produces a plasticized conducting polymer, which can be laminated into highly conductive and elastic films.[79]

Progress in maintaining the mechanical properties of the insulating host are also achieved in thermoplastic blends. Pioneering work on PVC/polypyrrole blends stated that flexible films were obtained, but stress-strain curves of the mixtures showed that the PVC mechanical behaviour was changed by adding the conducting polymer.[33] However, a mixture of high density polyethylene, HDPE, and PAni-DBSA showed that it is possible to retain the mechanical properties of the thermoplastic after the addition of an small concentration of PAni-DBSA.[3]

Highly conductive films are usually obtained in mixtures containing high contents of conducting polymers and it is hard task to maintain the mechanical properties of the insulating host with this addition. However, the mechanical behaviours of these mixtures are always superior to those observed for the pure conducting polymers. Besides, it was shown that the conducting polymer content needed to reach the percolation threshold to cause the onset of electronic conductivity is low. In addition, low contents of conducting polymers may be added to both elastomers and thermoplastics with no changes in the mechanical behaviour of the insulating hosts.

Thermal stability

An additional effect of adding a conducting polymer to an insulating matrix is the increase of the thermal stability of the latter. Large-scale methods for producing conductive blends involve processing at higher temperatures. Hence, the thermal stability of the components is crucial to obtain mixtures with high performances.

Thermogravimetric analysis of conductive blends based on PoAnis-TSA showed that, depending on the proportion between the components, the conducting polymer retards the degradation processes of the insulating matrix.[80] Figure 1 compares the thermal degradation process of several blends of this conducting polymer and poly(acrylonitrile-*co*-vinyl acetate), P(AN-VA), with theoretical curves estimated from the averages of the curves related to the pure polymers.

Peaks I, III and IV in Figure 1 are related to the degradation processes of the insulating matrix: (I) acetic acid liberation, (III) –CN scission and HCN liberation, (IV) cyclisation/crosslinking processes. In the mixtures, -CN scission of P(AN-VA) is less intense than expected (dotted curves), showing that the addition of PoAnis-TSA retards this process. This effect was attributed to the acid-acceptor behaviour of the conducting polymer. Probably, PoAnis-TSA interacts with the insulating matrix, via its non-protonated nitrogen atoms and dipoles formed in the P(AN-VA) chains, during the first step of the -CN scission process.

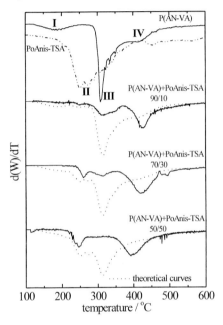

Figure 1 – 1st derivative TG curves of (−) pure P(AN-VA), (-..-..) pure PoAnis-TSA and blends containing different proportions between the components. Dotted lines correspond to the curves expected if no interaction occurs between the polymers. Copyright by Elsevier Science.[80]

Applications

Typical technological applications for conductive polymer blends include all kinds of systems based on pure conducting polymers. Small-scale electrochemical preparations are indicated for using in electrochemical devices, such as displays, smart windows, batteries and capacitors. In addition, large-scale preparation methods permit the production of conductive films with dimensions and mechanical properties adequate to be used in anti-static coatings, anticorrosive protection and electromagnetic shielding. Indeed, the use of conductive blends in some applications, instead of pure conducting polymers, leads to systems with optimised performances.

Conductive films for EMI shielding and corrosion protection

The increasing electromagnetic pollution, inherent in the use of electronic, computing and telecommunication equipment, and the not well-established effects of this electromagnetic interference (EMI) on human health are very relevant arguments in the modern World. Besides, perturbations from EMI also affect the operation of electronic systems, leading to malfunction or even failure of these devices.[81]

The approach of using conductive polymer blends, instead of inorganic-filled polymers, for this purpose is very interesting, because large amounts of metal/inorganic fillers are required to achieve reasonable EMI shielding, drastically affecting the mechanical properties of the mixtures. In fact, this was one of the first applications envisaged for conductive polymers.[82]

The first results of Taka[28], showing that blends of poly(3-octylthiophene) with several insulating thermoplastics (PVC, EVA and polystyrene) present reasonable EMI shielding effectiveness, encouraged the research in this field. For defence purposes, a large frequency range should be covered. Surface deposition of polypyrrole on PVC produced mixtures with microwave absorption from 0.1 to 20 GHz.[83] Mixtures of PAni and acrylonitrile-butadiene-styrene terpolymer, ABS, achieved shielding effectiveness of 60 dB at 101 GHz, with 50 wt% polyaniline.[84] The possibility of producing conductive films with suitable mechanical properties has stimulated research on the use of blends as EMI shielding materials and numerous examples can be found in the literature.[28,81-86]

Conducting polymers are also efficient to avoid metallic corrosion due to their unique redox properties. The mixture of these polymers with a suitable insulating matrix may lead to the production of very promising coatings for anti-corrosion purposes. The use of a mixture of PAni-CSA and epoxy resin to protect mild steel from an acid solution reduced the redox current by about five orders of magnitude and produced a 500 mV shift of the corrosion potential to the noble direction.[87] In another work, the same authors used blends based on both emeraldine base and HCl-doped PAni for the same purpose.[88]

Conductive films may also be used in a series of other protective coatings, such as marine antifouling agents[89] and electrostatic dissipation.[84]

Photolithography

Novel methods using photolithography to fabricate patterned electroactive films in highly specific microstructures are being extensively studied. Bidimensional structured patterns with micrometer and submicrometer spatial resolution can be achieved with this technique. Electroactive monomers that can be photopolymerised, and polymers that can be photocrosslinked or doped, can be used to fabricate microstructured electroactive polymer films adsorbed on optically transparent and conductive substrates.[90] The applications of such technology include individually addressable, pixel-sized devices that could be used to fabricate television-like screens.[91]

Some specific mixtures of conducting and insulating polymers have interesting properties for micrometer patternings. It is well known that PVC and other chlorinated polymers liberate hydrochloric acid under some conditions, such as UV, high energy irradiation or heating.[92] Otherwise, polyaniline changes its properties by acid doping/dedoping.

Emeraldine base blends with poly(vinylidene chloride-*co*-vinyl acetate) or PVC showed strong colour variations and conductivity increases by exposure to UV irradiation,[93] γ-radiation or e-beams.[94] These changes have been assigned to the dehydrochlorination of the polymer host and the doping of the emeraldine base by the HCl generated *in situ*. Taking into account that emeraldine base is significantly more soluble in several solvents than the parent HCl-doped PAni, this process can be used to produce lithographic patterns. Using a synchrotron source, Carinhana and co-workers irradiated a PVC/PAni base film and obtained patterns with micrometer resolution, as shown in Figure 2.[95]

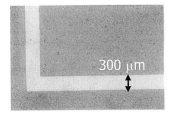 

Figure 2 - Lithographic patterns recorded with PAni-PVC films. Bright coloration corresponds to irradiation exposed areas. Copyright by John Wiley & Sons.[95]

96

Electrochromic devices

Electrochromic displays are another typical application for conducting polymers, in which better performances may be achieved if the pure polymer is replaced by a conductive polymer blend. For smart window applications, it is convenient to make flexible devices to permit coating of pre-existing windows. The use of conductive polymer blends with improved mechanical properties as electrochromic materials in such devices permits the construction of flexible systems, avoiding possible ruptures in the active films.[96]

Electrochromic properties and device stability are also optimised if a conductive blend is used. Figure 3 shows the optical contrast at 620 nm of the two elastomeric blends: (i) poly(ethylene oxide-*co*-epichlorohydrin)/poly(N,N-

dimethylbipyrrole),Poly(NNDMBP), and (ii) poly(ethylene oxide-*co*-epichlorohydrin)/poly(4,4'-dipentoxy-2,2'-bithiophene), poly(ET2), in a liquid electrolyte.[96]

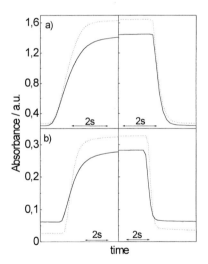

Figure 3 - Optical responses at 620 nm for colouring and bleaching steps. a) Poly(ET2), b) Poly(NNDMBP). Full lines: pure conducting polymers, dotted lines: blend materials. Copyright by Wiley-VCH.[96]

Both blends present higher chromatic contrast and faster colouring/bleaching kinetics than the pure conducting polymers. The elastomer was used in both blends because it is a good ionic conductor when mixed into inorganic salts.[97] The results were assigned to morphologic changes in the conducting polymer phase induced by the insulating host. Indeed, Amanokura and co-workers showed that interfacial transport is higher in a polypyrrole/polymer electrolyte blend than in the pure polypyrrole.[41]

Figure 4 compares the optical contrast at 620 nm for electrochromic devices based on these blends as a function of the colouring/bleaching cycles. It is clear that the device based on the elastomeric blends shows a longer lifetime than that based on the pure polymers. The presence of the insulating host probably precludes crosslinking reactions between the chains of the conducting polymer and improves the device performance.

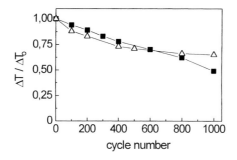

Figure 4 - Optical contrast of electrochromic devices at 620 nm as a function of the number of double potential steps: (△) using blend materials, (■) using pure poly(ET2) or poly(NNDMBP). Copyright by Wiley-VCH.[96]

Photovoltaic cells

The great demand for energy, associated with the limited resources of traditional fuels and environmental problems, has led to intensive research on the conversion of solar energy to electricity. Solar light is the most important source of energy and represents mankind's only inexhaustible energy source. One of the most promising tools to make use of solar energy is its direct conversion into electrical energy using photovoltaic or photoelectrochemical cells.[98]

These cells represent another good example of how conductive polymer blends may optimise the properties inherent to conducting polymers. Bulk donor-acceptor heterojunctions between conducting polymers and fullerenes have been used for photovoltaic devices with suitable energy conversion efficiency (around 0.04%).[99] By mixing a poly(*p*-phenylene-vinylene) derivative, PPV, a methane fullerene and polystyrene, Brabec and co-workers obtained an heterojunction with monochromatic power conversion efficiency of 1.5%.[100]

An improvement of the photoelectrochemical properties of PAni was observed by Neves and De Paoli for PAni supported in a microporous cellulose acetate (CA) membrane .[101] Figure 5 compares the photoelectrochemical responses of a pure PAni film and a CA/PAni film. It is noteworthy that the polymerisation of aniline in the pores of the cellulose acetate membrane intensifies the photoelectrochemical response of the polymer.

According to the authors, the insulating host permits a higher swelling of the photoactive material with the electrolyte solution containing a redox couple and favours the diffusion processes in the polymer | electrolyte interface. Also, the spongy-like structure of the membrane permits a deeper penetration of light, producing a more intense photoeffect.

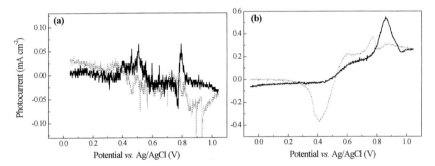

Figure 5 – Differential cyclic voltammetry for (a) pure PAni, (b) CA/PAni blend. Copyright by Elsevier Science.[101]

Light-emitting electrochemical cells

Light emitting diodes, LED, are the reverse of photovoltaic cells: while in the latter the light is used to produce an electric current, in LED a potential difference is applied to produce light. Taking into account the hole-injecting properties of the conducting polymers under an applied potential or current, flexible light emitting diodes based on these materials were first proposed by Burroughs and co-workers, generating great interest in the research of the so-called "polymer LED", also called OLED.[13]

A problem to be solved to achieve high efficiency in polymer LED is the preponderance of hole injection over electron injection, because metals with a sufficiently low work function necessary for the correct operation of these devices are chemically reactive. Besides, the faster mobility of holes in comparison to electrons also promotes a decrease in the device efficiency.

An alternative to avoid these difficulties are the *light-emitting electrochemical cells*, LEC's, first described by Heeger and co-workers in 1995.[102] LEC's are based on the characteristic of both p- and n-type doping of conducting polymers. A schematic representation of a LEC is shown in Figure 6, where the electroluminescent layer is composed of a blend of an electroluminescent polymer with a polymeric electrolyte. The electrolyte is needed to provide ions for the occurrence of both p- and n- doping in the polymer. Among the polymer electrolytes, the most used for LEC applications is the polymer complex composed by poly(ethylene oxide), PEO, and a lithium salt.[102,103]

By applying sufficiently high voltage between the metallic contacts, cations and anions move in opposite directions. The polymer side adjacent to the anode is oxidised and p-type carriers are introduced, while the polymer side closer to the cathode is reduced and n-type carriers are introduced. Electrochemical doping occurs when the applied potential is higher than the energy gap of the polymer. [104,105]

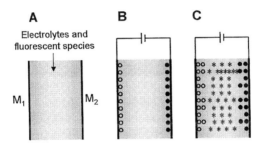

Figure 6 - Schematic representation of the electrochemical processes in a LEC. Electrodes are named M_1 and M_2, (o) is the oxidized species, (•) are the reduced species and (*) are the neutral electron-hole pairs. (A) The cell under zero bias, (B) Doping opposite sides as n- and p-types, (C) Charge migration and radiative decay. Copyright by AAAS.[102]

The charge balance in this process is provided by cation and anion mobility into the blend. As a consequence, a p-n junction is created in the bulk of the polymer film, where recombination processes and decay occur with light emission.

An interesting example of LEC was reported by Yang and Pei, who separately blended PPV and poly[(2-methoxy-5-(2'-ethylhexyloxy)-1,4-phenylene vinylene)], MEH-PPV, with an electrolyte consisting of PEO and a lithium salt.[106] The configuration of the electrochemical cell was Al | MEH-PPV blend | PPV blend | ITO and it showed red-orange emission under forward bias and green emission under reverse bias. According to the authors, a p-n junction is formed closer to the cathode due to the asymmetry of electron-hole injection and transport. Thus, the LEC emits red-orange light from the MEH-PPV layer when this side is charged negatively and emits green light from PPV when the PPV blend is charged negatively.

To enhance the performance of the LEC, Cao and co-workers included several liquid additives to the electroluminescent layer based on MEH-PPV + PEO + Li salt, to provide better compatibilisation between the components of the electroluminescent blend. [107] The surfactant-like behaviour of the additives causes a high surface area bicontinuous three-dimensional network morphology. The resulting LEC has better performance in comparison to the parent LEC without additive: the brightness increases from 10 to 1000 cd m^2. Besides, the additive-contaning LEC presents an efficiency of 2.5 %, which is similar to that obtained for a LED-type device using the same electroluminescent polymer.

Conclusions

The methods and results described indicate the variety of alternatives used to prepare conductive polymer blends. These mixtures have the typical applications envisaged for intrinsically conducting polymers. However, in some cases the resulting product has optimised properties in comparison to the pure conducting polymer. In addition, large-scale preparation methods (calendering, melt-mixing or extrusion) produce materials which can be used in other areas, such as EMI shielding.

The research on new conductive polymer blends has always aimed towards the production of processable materials. Although new low-cost preparation methods are still needed, large investments in this area reflect the interest and the optimism of chemists and technologists in obtaining new materials to fill several unexplored application niches.

Acknowledgements: The authors thank CNPq for research fellowships. WAG acknowledges the financial support from FAPERJ (E-26/170.096/2001) and CNPq (300149/2001-8). MAP acknowledges financial support from FAPESP (96/09983-0) and CNPq/PRONEx.

[1] H. Shirakawa, E. J. Louis, A. G. MacDiarmid, C. K. Chang, A.J. Heeger, *J. Chem. Soc., Chem. Commun.* **1977**, 578.

[2] S. Roth, W. Graupner, *Synth. Met.* **1993**, *57*, 3623.

[3] O. T. Ikkala, J. Laakso, E. Väkiparta, E. Virtanen, H. Ruohonen, H. Järvinen, T. Taka, P. Passiniemi, J. –E. Österholem, Y. Cao, A. Andreata, P. Smith, A. J. Heeger, *Synth. Met.* **1995**, *69*, 97.

[4] B. Wessling, *Synth. Met.* **1997**, *85*, 1313.

[5] F. Jonas, J. T. Morrison, *Synth. Met.* **1997**, *85*, 1397.

[6] S. Panero, P. Prosperi, F. Bonino, B. Scrosati, M. Mastragostino, *Electrochim. Acta* **1987**, *32*, 1007.

[7] C. Arbizzani, M. C. Gallazzi, M. Mastragostino, M. Rossi, F. Soavi, *Electrochem. Commun.* **2001**, *3*, 16.

[8] C. Arbizzani, M. Mastragostino, L. Meneghello, M. Morselli, A. Zanelli, *J. Appl Electrochem.* **1995**, *26*, 121.

[9] W. A. Gazotti, G. Casalbore-Miceli, A. Geri, M. –A. De Paoli, *Adv. Mater.* **1998**, *10*, 60.

[10] W. A. Gazotti, M. -A. De Paoli, G. Casalbore-Miceli, A. Geri, G. Zotti, *J. Appl. Electrochem.* **1999**, *29*, 753.

[11] C. T. Kuo, W. H. Chiou, *Synth. Met.* **1997**, *88*, 23.

[12] H. Sirringhaus, N. Tessler, R. H. Friend, *Synth. Met.* **1999**, *102*, 857.

[13] J. H. Burroughes, D. D. C. Bradley, A. R. Brown, R. N. Marks, K. Mackay, R. H. Friend, P. L. Burns, A. B. Holmes, *Nature* **1990**, *347*, 539.

[14] N. Camaioni, G. Casalbore-Miceli, A. Geri, G. Zotti, *J. Phys. D: Appl. Phys.* **1998**, *31*, 1245.

[15] T. Yohannes, O. Inganäs, *Sol. Energy Mater. Sol. Cells* **1998**, *51*, 193.

[16] W. A. Gazotti, A. F. Nogueira, E. M. Girotto, M. -A. De Paoli, *Synth. Met.* **2000**, *108*, 151.

[17] W. A. Gazotti, E. M. Girotto, A. F. Nogueira, M. -A. De Paoli, *Sol. Energy Mater. Sol. Cells* **2001**, *69*, 315.

[18] A. Prón, J. Laska, J. E. Österholm, P. Smith, *Polymer* **1993**, *34*, 4235.

[19] A. Prón, J. E. Österholm, P. Smith, A. J. Heeger, J. Laska, M. Zagórska, *Synth. Met.* **1993**, *55-57*, 3520.

[20] J. Laska, A. Prón, M. Zagórska, S. Lapkowski, S. Lefrant, *Synth. Met.* **1995**, *69*, 113.

[21] J. Laska, P. Izak, P. A. Prón, *J. Appl. Polym. Sci.* **1996**, *61*, 1339.

[22] Y. Cao, P. Smith, A. J. Heeger, *Synth. Met.* **1992**, *48*, 91.

[23] G. Beyer, B. Steckenbiegler, *Synth. Met.* **1993**, *60*, 169.

[24] A. J. Heeger, *Synth. Met.* **1993**, *55-57*, 3471.

[25] J. P. Pouget, C. –H. Hsu, A. G. MacDiarmid, A. J. Epstein, *Synth. Met.* **1995**, *69*, 119.

[26] M. –A. De Paoli, in: *"Handbook of Organic Conductive Molecules and Polymers"*, H. S. Nalwa, Ed., John Wiley & Sons, New York 1997, p. 773.

[27] A. Echte, *"Handbuch der Technischen Polymerchemie"*, VCH Verlag, Weinheim 1993, p.663.

[28] T. Taka, *Synth. Met.* **1991**, *41*, 1177.

[29] M. –A. De Paoli, R. A. Zoppi, M. I. Felisberti, in: *"Polymeric Materials Encyclopedia"*, J. C. Salamone, Ed., CRC Press, Boca Raton 1996, p.1455.

[30] M. C. Jesus, Y. Fu, R. A. Weiss, *Polym. Eng. Sci.* **1997**, *37*, 1936.

[31] J. Anand, S. Palaniappan, D. N. Sathyanarayana, *Prog. Polym. Sci.* **1998**, *23*, 993.

[32] M. –A. De Paoli, R. J. Waltman, A. F. Diaz, J. Bargon, *J. Chem. Soc., Chem. Commun.* **1984**, 1015.

[33] M. –A. De Paoli, R. J. Waltman, A. F. Diaz, J. Bargon, *J. Polym. Sci., Polym. Chem.* **1985**, *23*, 1687.

[34] O. Niwa, M. Hikita, T. Tamamura, *Appl. Phys. Lett.* **1985**, *46*, 444.

[35] J. Roncali, F. Garnier, *J. Chem. Soc., Chem. Commun* **1986**, 783.

[36] G. Casalbore-Miceli, G. Beggiato, A. Geri, A. Berlin, G. Pagani, *Mol. Cryst. Liq. Cryst.* **1992**, *219*, 173.

[37] M. –A. De Paoli, D. J. Maia, *J. Mater. Chem.* **1994**, *4*, 1799.

[38] C. M. Chang, H. J. Huang, *Anal. Chim. Acta* **1995**, *300*, 15.

[39] Y. H. Park, C. R. Park, *Synth. Met.* **2001**, *118*, 187.

[40] H. Naarman, in: *"Science and Applications of Conducting Polymers"*, W. R. Salaneck, D. T. Clark, E. J. Samuelsen, Ed., Adam Hilger, Bristol 1990.

[41] J. Amanokura, Y. Suzuki, S. Imabayashi, M. Watanabe, *J. Electrochem. Soc.* **2001**, *148*, D43.

[42] M. E. Nicho, H. Hu, *Sol. Energy Mater. Sol. Cells* **2000**, *63*, 423.

[43] R. A. Zoppi, M. I. Felisberti, M. –A. De Paoli, *J. Polym. Sci, Polym. Chem.* **1994**, *32*, 1001.

[44] V. Mano, M. I. Felisberti, M. –A. De Paoli, *Polymer* **1996**, *37*, 5170.

[45] M. Wan, M. Li, J. Li, Z. Liu, *Thin Solid Films* **1995**, *259*, 188.

[46] H. S. Lee, J. Hong, *Synth. Met.* **2000**, *113*, 115.

[47] L. H.C. Mattoso, R. M. Faria, L. O. S. Bulhões, A. G. MacDiarmid, *J. Polym. Sci., Polym. Chem.*

1894, *32*, 2147.

[48] W. A. Gazotti, M. –A. De Paoli, *Synth. Met.* **1996**, *80*, 263.

[49] D. Gonçalves, D. S. dos Santos, L. H. C. Mattoso, F. E. Karasz, L. Alcekud, R. M. Faria, *Synth. Met.* **1997**, *90*, 5.

[50] L. F. Malmonge, L. H. C. Mattoso, *Polymer* **1995**, *36*,245.

[51] C. Y. Yang, M. Reghu, A. J. Heeger, Y. Cao, *Synth. Met.* **1996**, *79*, 27.

[52] W. Lee, Y. Kim, S. Kaang, *Synth. Met.* **2000**, *113*, 237.

[53] F. Cataldo, P. Maltese, *Polym. Adv. Techn.* **2001**, *12*, 293.

[54] I. D. Norris, M. M. Shaker, F. K. Ko, A. G. MacDiarmid, *Synth. Met.* **2000**, *114*, 109.

[55] D. Gonçalvez, A. Waddon, F. E.Karasz, L. Alcekud, *Synth. Met.* **1995**, *74*, 197.

[56] W. A. Gazotti, G. Casalbore-Miceli, S. Mitzakoff, A. Geri, M. C. Gallazzi, M. –A. De Paoli, *Electrochim. Acta* **1999**, *44*, 1965.

[57] W. A. Gazotti, R. Faez, M. –A. De Paoli, *Eur. Polym. J.* **1999**, *35*, 35.

[58] M. C. Su, J. L. Hong, *Polymer* **2001**, *42*, 3297.

[59] S. J. Davies, T. G. Ryan, C. J. Wilde, G. Beyer, *Synth. Met.* **1995**, *69*, 209.

[60] S. Mitzakoff, M. –A. De Paoli, *Eur. Polym. J.* **1999**, *35*, 1791.

[61] R. K. Paul, C. K. S. Pillai, *Synth. Met.***2000**, *114*, 27.

[62] M. R. Vallim, M. I. Felisberti, M. –A. De Paoli, *J. Appl. Polym. Sci.* **2000**, *75*, 677.

[63] R. Faez, W. A. Gazotti, *Polymer* **1999**, *40*, 5497.

[64] R. Faez, M. –A. De Paoli, *Eur. Polym. J.* **2001**, *37*, 1139.

[65] Y. Haba, E. Segal, M. Narkis, G. I. Titelman, A. Siegmann, *Synth. Met.* **2000**, *110*, 189.

[66] N. Gospodinova, P. Mokreva, T. Tsanov, L. Terlemezyan, *Polymer* **1997**, *38*, 743.

[67] R. Zallen, *"The Physics of Amorphous Solids"*, John Wiley & Sons, New York 1983.

[68] J. P.Yang, P. Rannou. J. Planes, A. Prón, M. Nechtschein, *Synth. Met.* **1998**, *93*,169.

[69] M. Narkis, Y. Haba, E. Segal, M. Zilberman, G. I. Titelman, A. Siegman, *Polym. Adv. Tech.* **2000**, *11*, 665.

[70] R. C. Patil, K. Kuratani, M. Kakayama, K. Ogura, *J. Polym. Sci., Polym. Chem.* **1999**, *37*, 2657.

[71] C. O. Yoon, M. Reghu, D. Moses, Y. Cao, A. J. Heeger, *Synth. Met.* **1995**, *69*, 255.

[72] P. Juvin, M. Hasik, J. Fraisse, J. Planes, A. Prón, I. Kulszewicz-Bajer, *J. Appl. Polym. Sci.* **1999**, *74*, 471.

[73] J. Fraisse, J. Planes, A. Dufresne, *Mol. Cryst. Liq. Cryst.* **2000**, *354*, 1099.

[74] S. M. Yang, H. L. Lee, *Synth. Met.* **1999**, *102*, 1226.

[75] G. M. O. Barra, M. E. Leyva, B. G. Soares, L. H. Mattoso, M. Sens, *J. Appl. Polym. Sci.* **2001**, *82*, 114.

[76] T. Makela, T. Haatainen, J. Ahopelto, H. Isolato, *Synth. Met.* **2001**, *121*, 1309.

[77] A. R. Hopkins, J. R. Reynolds, *Macromolecules* **2000**, *33*, 5221.

[78] M. G. Han, S. S. Im, *J. Appl. Polym. Sci.* **1998**, *67*, 1209.

[79] R. Faez, M. –A. De Paoli, *J. Appl. Polym. Sci.* **2001**, 82, 1768.

[80] W. A. Gazotti, P. S. Freitas, W. R. Waldman, M. –A. De Paoli, *Synth. Met.* **1999**, *102*, 1207.

[81] J. A. Pomposo, J. Rodríguez, H. Grande, *Synth. Met.* **1999**, *104*, 107.

[82] H. J. Mair, D. S. Roth, *"Elektrisch Leitende Kunststoffe"*, Technische Akademie Esslingen Symposium, Esslingen 1985.

[83] L. Olmedo, P. Hourquebie, F. Jousse, *Adv. Mater.* **1993**, *5*, 373.

[84] S. Koul, R. Chandra, S. K. Dhavan, *Polymer* **2000**, *41*, 9305.

[85] S. Courric, V. H. Tran, *Polym. Adv. Tech.* **2000**, *11*, 273.

[86] R. Faez, I. M. Martin, M. –A. De Paoli, M. C. Rezende, *Synth. Met.* **2001**, *119*, 435.

[87] A. Talo, P. Passiniemi, O. Forsén, S. Yläsaari, *Synth. Met.* **1997**, *85*, 1333.

[88] A. Talo, O. Forsen, S. Yläsaari, *Synth. Met.* **1999**, *102*, 1394.

[89] X. H. Wang, J. Li, J. Y. Zhang, Z. C. Sun, L. Yu, X. B. Jing, F. S. Wang, Z. X. Sun, Z. J. Ye, *Synth. Met.* **1999**, *102*, 1377.

[90] T. S. Bergstedt, B. T. Hauser, K. S. Schanze*, J. Am. Chem. Soc.* **1984**, *116*, 8380.

[91] M. L. Renak, G. C. Bazan, D. Roitman, *Synth. Met.* **1998**, *97*, 17.

[92] E.D. Owen, *"Degradation and Stabilisation of PVC"*, Elsevier Applied, Essex 1984.

[93] H. Bodugoz, U. A. Sevil, O. Güven, *Macromol. Symp.* **2001**, *169*, 289.

[94] U. A. Sevil, O. Güven, Ö. Birer, S. Süzer, *Synth. Met.* **2000**, *110*, 175.

[95] D. Carinhana, M. –A. de Paoli, C. S. C. Castro, G. G. B. Souza, *Adv. Mater. Opt. Electron.* **2000**, *10*, 241.

[96] W. A. Gazotti, G. Casalbore-Miceli, A. Geri, A. Berlin, M. -A. De Paoli; *Adv. Mater.*, **1998**, *10*, 1522.

[97] W. A. Gazotti, M. A. S. Spinacé, E. M. Girotto, M. -A. De Paoli; *Solid State Ionics*, **2000**, *130*, 281.

[98] W. A. Gazotti, A. F. Nogueira, E. M. Girotto, L. Micaroni, M. Martini, S. Neves, M. -A. De Paoli, in: *"Handbook of Advanced Electronic and Photonic Materials and Devices"*; H. S. Nalwa, Ed., Academic Press, San Diego 2001, v. 10, p. 54.

[99] C.J. Brabec, N.S. Sariciftci, *Mater. Today* **2000**, 5.

[100] C. J. Brabec, F. Padinger, N. S. Saricifti, J. C. Hummelen, *J. Appl Phys.* **1999**, *85*, 6866.

[101] S. Neves, M.-A. De Paoli, *Synth. Met.* **1998**, *96*, 49.

[102] Q. Pei, G. Yu, C. Zhang, Y. Yang, A. J. Heeger, *Science* **1995**, *269*, 1086.

[103] Q. Pei, Y. Yang, G. Yu, Y. Cao, A. J. Heeger, *Synth. Met.* **1997**, *85*, 1229.

[104] G. Yu, Y. Yang, Y. Cao, Q. Pei, C. Zhang, A. J. Heeger, *Chem. Phys. Lett.* **1996**, *259*, 465.

[105] D. J. Dick, A. J. Heeger, Y. Yang, Q. Pei, *Adv. Mater.* **1998**, *8*, 985.

[106] Y. Yang, Q. Pei, *Appl. Phys. Lett.* **1996**, *68*, 2708.

[107] Y. Cao, G. Yu, A. J. Heeger, C. Y. Yang, *Appl. Phys. Lett.* **1996**, *68*, 23.

Shear Controlled Morphology of Rubber/Organoclay

Nanocomposites and Dynamic Mechanical Analysis

Frank Schön, Ralf Thomann, Wolfram Gronski*

Universität Freiburg, Institut für Makromolekulare Chemie,
Stefan-Meier-Str. 31, 79104 Freiburg, Germany

Summary: Styrene butadiene rubber/layered silicate composites were prepared using clays modified with alkyl ammonium salts of different structure. The influence of modifier structure and compounding conditions on the extent of exfoliation/intercalation of the silicates was investigated by electron microscopy and dynamic mechanical analysis. With increasing exfoliation the glass transition temperature is lowered and the stability of the filler network is enhanced.

Introduction

In the last few years layered silicates have attracted much interest as fillers with potential for novel applications. The clays are modified via cation exchange with ammonium salts to yield organophilic clays (OC) and to make them compatible with polymers. During mixing the polymer chains can penetrate into the gallery between the layers and under the action of shear forces the aggregates can be transformed into smaller intercalated stacks of layers or even exfoliated to single layers leading to polymer/silicate nanocomposites (NC). Due to the ultrafine dispersion and high aspect ratio of the OC the NC exhibit superior strength, moduli, higher heat distortion temperatures and improved water and gas barrier properties compared to conventional isotropic fillers as silica or carbon black, especially when the layers are exfoliated into single layers.[1, 2] A large number of papers deals with mechanical properties of NC with polyamide[3], polypropylene[4], epoxy resins[5] and rubber[6,7]. This work focuses on the generation of styrene butadiene rubber (SBR)/Organoclay composites with different types of OC and investigations on the influence of the filler dispersion on dynamic mechanical relaxation. The T_g dependence and a relaxation process due to filler network destruction are investigated in terms of exfoliation degree. The influence of shear forces during compounding on the filler dispersion is also discussed.

       CCC 1022-1360/00/$ 17.50+.50/0

Experimental

To prepare nanocomposites with SBR (BUNA SL 18-0, 18% styrene content, $M_w \approx 400.000$ g/mol, Bayer AG) as matrix polymer three different organically modified clays were used. The modification takes place via a exchange of the intergallery cations of the layered silicate against ammonium cations and results in OC_1, OC_2 (Nanofil, Südchemie AG) and OC_3 (Nanomer I.42E, Nanocor). The clay used for the preparation of OC_1[6] is a synthetic hectorite (Somasif ME 100, Coop Chemical), OC_2 and OC_3 are based on montmorollinite. The only difference of the latter ones is the intergallery distance determined by X-ray scattering, OC_1 has a spacing of 2.7 nm between two silicate layers, OC_2 3.1 nm and in the case of OC_3 it is 3.6 nm. The particle size of the OC is about 10-25 µm.

For compounding with SBR an internal mixer (Haake Rheocord 90, Banbury rotors) was used. SBR, vulcanization chemicals (1.5 phr sulfur, 2,5 phr N-Cyclohexyl-2-benzothiazole sulfene amide (CBS), Bayer AG, 4 phr zinc oxide, 2 phr stearic acid) and the organoclay (30 phr, which is equal to 15 wt% inorganic content) were mixed at temperatures of 80-90°C for 10 minutes. For further improvement of filler dispersion the mixing was followed by a treatment on a roll mill at 80°C for 10 min. After compounding the products were vulcanized for 50 min at 160° C in a hot stage press (Collin).

Dynamic mechanical analysis (DMA) was performed on a Rheometrics RSA II in film fixture geometry (1 mm x 6 mm x 22 mm) at a frequency of 1Hz and 0.1 % strain from -80 °C to 120 °C.

Transmission electron micrographs were taken from ultrathin sections cryo-ultramicrotomed at -100 °C with a Reichert Ultracut E using a LEO 912 Omega operating at a voltage of 120 kV and a slow scan camera.

Influence of organoclay structure on intercalation/exfoliation

Vaia et al. described a lattice model of polymer melt intercalation in organoclays where they separated the free energy change upon mixing into an energy change due to new intermolecular interactions and a combinatorial entropy change associated with conformational changes of the polymer and the modifier.[8] When the polymer diffuses into the galleries between the silicate layers it loses conformational freedom due to confinement between the layers which can be compensated partly by enhanced conformational freedom of the modifier chains and more efficiently by the energy gain through favourable interactions between polymer/modifier and polymer/silicate. If the energy gain is smaller than the entropic loss the polymer chains do not diffuse into the galleries. In the case of favourable enthalpic interactions intercalation and exfoliation under the combined action of thermodynamic and shear forces is possible. The morphology image of the composite containing OC_1 shows that intercalation of the polymer is prevented (figure 1), whereas nanocomposites with intercalated and exfoliated structures are generated with OC_2 and OC_3 (figure 2). The favoured diffusion into the OC galleries in the second case probably occurs because of a larger initial entropy gain of the modifier due to the longer alkyl chains. In the first case the entropy barrier opposing diffusion cannot be compensated by the assumed favourable interactions between the phenolic group of OC_1 and the phenyl ring of SBR.

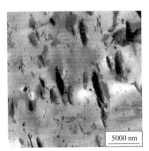

Figure 1: large silicate
aggregates (OC_1) in SBR

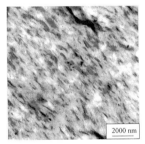

Figure 2: well dispersed
silicate layers (OC_3) in SBR

Influence of shear forces on the extent of filler exfoliation

When high shear forces are applied during compounding, as it is the case on a roll mill, the shear stress supports the breakdown of larger silicate stacks into smaller ones. If the thermodynamic factors discussed above are favourable the shear stress and polymer

108

diffusion into the galleries force the silicate layers to peel apart one by one. Fornes et al. recently proposed this mechanism for exfoliation under shear[9]. It is schematically shown in figure 3. A fan-shaped intermediate observed by TEM (figure 4) visualizes this mechanism.

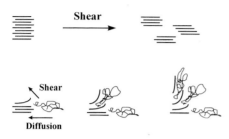

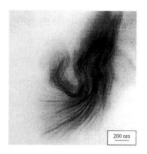

Figure 3: Schematic mechanism for exfoliation under shear[9]

Figure 4: Silicate layers peel apart via a fan-shaped intermediate

The breakdown of large silicate aggregates (tactoids) into smaller stacks by virtue of greater shear forces can also be visualized by TEM images at different compounding stages. Figure 5 shows an image after compounding in an internal mixer where moderate shear forces are applied. After treatment on a roll mill under high shear stress large agglomerates are broken into smaller stacks (figure 6).

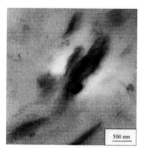

Figure 5: OC$_2$ in SBR matrix after compounding on an internal mixer

Figure 6: OC$_2$ in SBR matrix after compounding on a roll mill

Influence of filler dispersion on dynamic mechanical relaxation

First it is observed that the glass transition is shifted to lower temperature as the degree of intercalation/exfoliation increases (figure 7). When the silicate layers are exfoliated a greater amount of ammonium salt is at the silicate surface and acts as a plasticizer. The decrease of the tan δ peak with enhanced degree of intercalation/exfoliation is due to the greater amount of polymer included in intercalated stacks. This reduces the amount of polymer being deformed during strain oscillation and thus reduces the amount of dissipated energy.

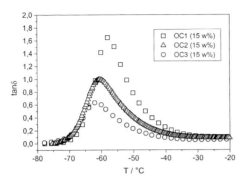

Figure 7: T_g dependence on filler dispersion

The additional relaxation process at higher temperatures can be related to the process associated with the breakdown and reformation of the filler network. The process is shifted to higher temperatures with increasing degree of exfoliation showing the greater perfection and stability of the network containing a larger number of exfoliated layers (figures 8 - 9). In conventional silica filled systems a much higher filler content is necessary to observe this relaxation process[10].

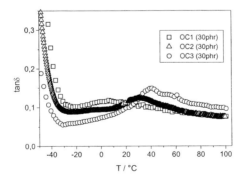

Figure 8: Temperature dependence of filler network relaxation on filler exfoliation

Both E' and tan δ show that exfoliation and expression of the filler network are increased with organoclay structure from $OC_1 \rightarrow OC_2 \rightarrow OC_3$.

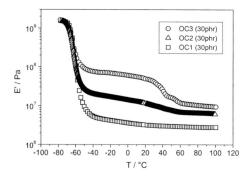

Figure 9: Temperature dependence of E' for organoclays with different degree of exfoliation

Conclusions

Introduction of a group with favourable modifier/polymer interactions into the modifier may not be sufficient for polymer intercalation if the length of the alkyl chain of the modifier is not long enough. If entropic and enthalpic interactions are favourable it is possible to improve exfoliation by increasing the shear stress during compounding. Exfoliation is shown to proceed via fan-shaped intermediates by TEM. DMA was proven to be a sensitive method to monitor the degree of exfoliation and intercalation by the depression of T_g and by the intensity and temperature dependence of the loss process at higher temperature originating from the relaxation of the filler network.

References

[1] Giannelis, E. P. *Adv. Mater.* **1996**, *8*, 29.
[2] Alexandre, M., Duois, P. *Mat. sci. eng.* **2000**, *28*, 1.
[3] Reichert, P., Kressler, J., Thomann, R., Mülhaupt, R., Stöppelmann, G. *Acta Polymerica* **1998**, *49*, 116
[4] Reichert, P., Nitz, H., Klinke, S., Brandsch, R., Thomann, R., Mülhaupt, R. *Macrom. Mater. Eng.* **2000**, *275*, 8.
[5] Zilg, C., Mülhaupt, R., Finter, J. *Macromol. Chem. Phys.* **1999**, *200*, 661.
[6] Ganter, M., Gronski, W., Semke, H., Zilg, C., Thomann, R. Mülhaupt, R. *KGK Kautschuk Gummi Kunststoffe* **2001**, *54*, 166.
[7] Ganter, M., Gronski, W., Reichert, P., Mülhaupt, R. Rubber Chem. Technol. **2001**, 74, 221
[8] Vaia, R. A., Giannelis, E.P. *Macromolecules* **1997**, *30*, 7990.
[9] Fornes, T. D., Yoon, P.J., Paul, D.R. *Polymer* **2001**, *42*, 9929.
[10] Wang, M.-J. *Rubber Chem. Technol.* **1999**, *72*, 430.

Macromol. Symp. **189**, *111–125 (2002)*

Results Coming from Homogeneous and Supported Metallocene Catalysts in the Homo- and Copolymerization of Olefins

Raúl Quijada[1], Jaime Retuert[1], Juan L. Guevara[1], René Rojas[1],
Marcela Valle[1], Pilar Saavedra[1], Humberto Palza[1], Griselda B. Galland[2]

[1]Departamento de Ingeniería Química, Facultad de Ciencias Físicas y Matemáticas, Universidad de Chile, Casilla 2777, Santiago, Chile, and Centro para la Investigación Multidisciplinaria Avanzada en Ciencias de los Materiales
[2]Instituto de Quimica, Universidade Federal do Rio Grande do Sul, Av. Bento Gonçalves 9500-91501-970 Porto Alegre-RS-Brasil

Summary: Ethylene, propylene and α-olefins were homo- and copolymerized in the presence of a series of homogeneous catalytic systems consisting of methylaluminoxane (MAO) and group IV metallocenes such as $Et(Ind)_2ZrCl_2$ (I), $Me_2Si(Ind)_2ZrCl_2$ (II), $Et(2\text{-}Me\text{-}Ind)_2ZrCl_2$ (III), $Ph_2C(Flu)(Cp)ZrCl_2$ (IV). It was found that the catalytic activity, the incorporation of comonomer in the case of copolymers, and the microstructure of the polymers depend on the catalyst's structure. For heterogeneous catalysts, several supports based on metal oxide compounds have been investigated, with special emphasis in those obtained by the sol-gel preparation technique. The homo- and copolymerization of the monomers in the homogeneous systems studied where also investigated using the same catalyst system, but in a heterogeneous medium. Comparative results from the homogeneous and heterogeneous systems are presented and discussed.

Introduction

Homopolymers and copolymers of ethylene and/or propylene with long-chain α-olefins obtained via catalysis are important commercial materials, since their development for different applications has been characterised by their major production and consumption in the last few years, and consequently great efforts have been made to find new and more efficient catalysts to produce these polymers[1].

Advances in the design of single site catalysts are providing an ever-growing menu of metal-ligand combinations suitable for catalysing the polymerization of olefins[2]. These catalysts enable control over the polymerization reaction and therefore of important bulk properties of the resulting polymer[3]. An excellent body of mechanistic work also exists that provides insight into how the active site technology is making a considerable impact on the commercial processes that produce these commodity products[4-5].

   CCC 1022-1360/00/$ 17.50+.50/0

The use of metallocene catalysts has allowed a very rapid development in the field of polyolefins. These catalysts present single-site characteristics (and very high activities), and thus all the sites produce nearly the same chain architecture, leading to polymers with a narrow molecular weight distribution. In the case of copolymers with α-olefins, the side branches are randomly distributed in the polymer backbone.[6-8]

Considering the extensive experience acquired by our group in the study of the homo- and copolymerization of ethylene with several α-olefins using metallocene catalysts, we have found it convenient to study the coordinating action of different pairs of catalysts for making polyethylene copolymers from a single readily available monomer, and to explore the possibility of producing these materials with elastomeric properties. The homo- and copolymerization of propylene with α-olefins, and the preparation of heterogeneous catalysts based on metallocene compounds were also studied.

In order to study the effect of the metallocene structures on the polymerization of olefins we chose four metallocenes (Figure 1): Et(Ind)$_2$ZrCl$_2$ (I), Me$_2$Si(Ind)$_2$ZrCl$_2$ (II), Et(2-Me-Ind)$_2$ZrCl$_2$ (III), Ph$_2$C(Flu)(Cp)ZrCl$_2$ (IV).

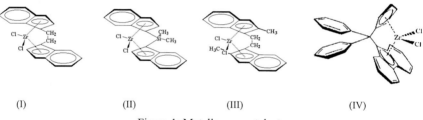

(I)   (II)   (III)   (IV)

Figure 1: Metallocene catalysts

With these catalyst systems it is possible to analyse the influence of the different types of interannular bridges (-CH$_2$-CH$_2$-; (CH$_3$)$_2$Si), the presence of substituents in the aromatic ring (catalysts I and III), and finally the type of stereoselectivity that the catalyst has on the microstructure and properties of the resulting polymer.

Homogeneous Catalysts

Our earlier studies[9-13] showed that several metallocene catalysts promote the copolymerization of ethylene with α-olefins. Copolymers of ethylene-1-hexene and ethylene-1-octadecene with highly elastomeric properties were obtained. Activities around 50,000 [kg polymer/ mol h bar] for the homopolymerization of ethylene and 100,000 [kg polymer/ mol h bar] for copolymerization were obtained (Figure 2) and incorporation values of up to 18 mol-% for the 1-hexene comonomer (Figure 3).

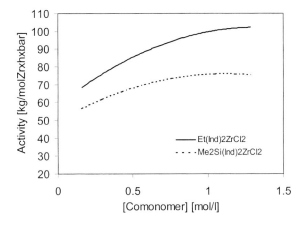

Figure 2: Influence of 1-hexene concentration on catalytic activity for two different metallocene systems.

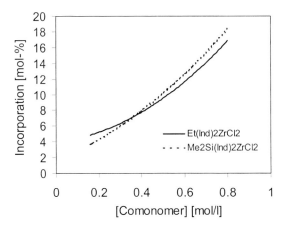

Figure 3: Influence of 1-hexene concentration on comonomer incorporation for two different metallocene systems.

The Et(Ind)$_2$ZrCl$_2$/MAO catalytic system in both cases (homo- and copolymerization) showed the highest catalytic activity (Figures 2 and 4). The Me$_2$Si(Ind)$_2$ZrCl$_2$/MAO (II) system, however, reaches a large comonomer incorporation in the copolymerization of ethylene with α-olefins (Figures 3 and 5), especially with long-chain olefins like 1-octadecene,[9-10] as shown in Figure 5.

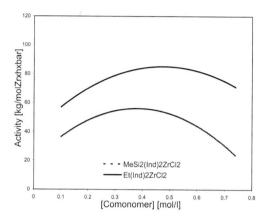

Figure 4: Influence of 1-octadecene concentration on catalytic activity for two different metallocene systems.

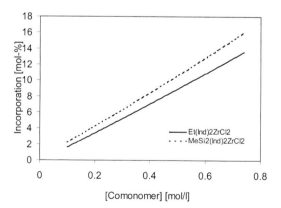

Figure 5: Influence of 1-octadecene concentration on comonomer incorporation for two different metallocene systems.

Results obtained from the stress-strain behaviour of some ethylene-1-octadecene copolymers, obtained and quenched from the melt, are shown in Figure 6. It is clear that the effect of the incorporated comonomer changes the mechanical behaviour of the curve, going from a rigid material as PE to one elastomeric type material in the case of C4 (highest incorporation).[11]

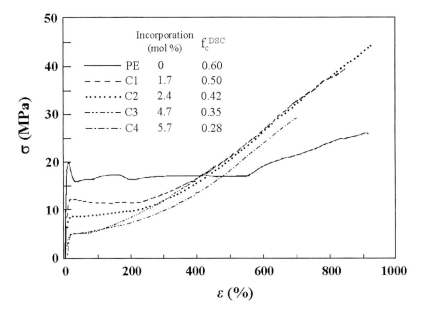

Figure 6: Stress-strain curves for different samples at room temperature

For the homopolymerization of propylene, the trends found are the same as for ethylene, which mean that the highest activity values were found with the $Et(Ind)_2ZrCl_2/MAO$ (I) catalytic system[12].

For the copolymerization of propylene with different comonomers, the results obtained with catalytic system I show high catalytic activities. Figure 7 shows the results of these studies. An increase in activity is seen for all the comonomers.

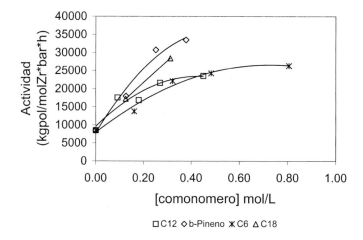

Figure 7: Influence of comonomer concentration on catalytic activity for propylene copolymerization.

The study related to the homopolymerization of long-chain α-olefins is also quite useful, since very valuable information can be obtained that makes it possible to understand more completely the behavior of α-olefins when used as comonomers in copolymerization reactions with other olefins[13-14]. The homopolymerization of 1-octadecene by using rac-Et(Ind)$_2$ZrCl$_2$/MAO (I) and rac-Me$_2$Si(Ind)$_2$ZrCl$_2$/MAO (II) as catalyst systems has already been reported by our research group. Table 1 shows the main results.

Table 1: Influence of temperature and catalytic system on activity and molecular weight in the homopolymerization of 1-ocatadecene.

Catalyst	Temperature °C	Conversion %	Activity [kg Pol / mol Zr bar]	Mw [g/mol]	Mw/Mn
I	70	77	1200	6326	1.6
	30	28	470	22647	1.8
II	70	33	530	9943	1.5
	30	25	400	25044	1.7
III	70	63	1000	5473	1.6
	30	25	400	22055	1.7
IV	70	47	730	19056	1.6
	30	7	110	37676	2.2

The results show that the catalytic activity of system III is similar to that obtained for system I, where the methyl group does not have any effect. The catalytic activity of system II was found to be the lowest of all the catalyst systems studied. This behavior is possibly associated with charge effects on the metallic center, and may also be due to the opening of the angle between the indenyl rings which leads to a change in the bridge.

This also shows that the catalytic activities of all the catalyst systems decreased when the polymerizations were performed at 30° C, giving similar activity values for catalyst systems I, II and III. Catalyst system IV showed the lowest catalytic activity at 30° C. The catalytic activity of system IV was 6.5 times lower for polymerizations carried out at 30° C as compared with those performed at 70° C[15].

Finally, we have looked into the importance of the copolymers of ethylene with α-olefins. In this context we have also studied this copolymerization producing the comonomer in situ via oligomerization of ethylene with an iron catalyst [((2-ArN-C(Me))$_2$C$_5$H$_3$)FeCl$_2$] (V). The polymers obtained with this system have different branch lengths incorporated in the main polymer chain, as shown in Figure 8.[16]

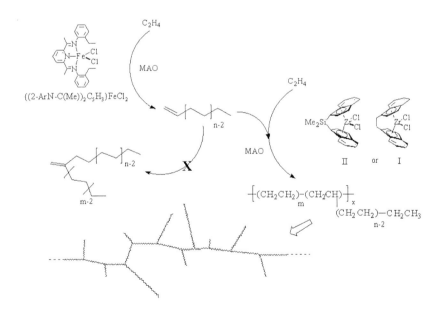

Figure 8: Simultaneous oligomerization and polymerization of ethylene with iron and zirconium catalysts.

Table 2: Ethylene copolymerization by producing the comonomer in situ: Fe/EtInd$_2$ZrCl$_2$/MAO catalytic system[a].

Entry	Fe/Zr	Activity[b]	T_m (°C) [c]	M_w x 10^{-3}	M_w/M_n	Incorporation mol % [d]
1	4.6	34,060	124	81.2	1.68	2
2	9.3	16,850	112	65.1	1.65	2.8
3	1.9	27,884	119	74.6	1.56	1.8
4	3.1	18,259	119	81.8	1.61	2.2
5	4.7	16,180	117	68.7	1.68	2.5
6	2	20,850	121	81.1	1.91	1.8

Table 3: Ethylene copolymerization by producing the comonomer in situ: Fe/Me$_2$Si(Ind)$_2$ZrCl$_2$/MAO catalytic system[a].

Entry	Fe/Zr	Activity[b]	T_m (°C) [c]	M_w x 10^{-3}	M_w/M_n	Incorporation mol % [d]
1	4.5	20,933	123	55.9	16.1	2.4
2	10	9,850	125	29.9	13.2	3
3	2.9	19,455	121	83.9	15.6	2.9
4	4.5	12,090	124	57.3	15.2	4
5	2	12,675	120	80.5	9.1	2.3

For both tables:

(a) Reaction Conditions: Solvent, toluene; polymerization time, 0.5 h.; stirrer speed, 500 rpm; pressure, 2 bar; temperature, 60 °C; Al/Fe, 2000; Al/Zr, 2000-15000. **(b)** kg polymer/(molZr+molFe)(h)(bar). **(c)** Melting Point determined by DSC. **(d)** Moles of branching per 100 ethylene units.

A summary of the results obtained for the different Fe/EtInd$_2$ZrCl$_2$/MAO/C$_2$H$_4$ and Fe/Me$_2$Si(Ind)$_2$ZrCl$_2$/MAO/C$_2$H$_4$ combinations are given in Tables 2 and 3, respectively. Analysing Table 2, it was found that for a given concentration of Zr, increasing the Fe/Zr ratio leads to a reduction in the activity. Relative to Fe/Me$_2$Si(Ind)$_2$ZrCl$_2$/MAO/C$_2$H$_4$ combinations, the same trend was observed for this system with respect to the catalyst content, which means that the increase in the Fe/Zr ratio decreases the activity, and when the total metal concentration is raised, a lower activity is observed. On the other hand, a larger amount of ethylene was consumed by the Fe/EtInd$_2$ZrCl$_2$/MAO/C$_2$H$_4$ catalyst system. It was also observed that for a constant Fe/Zr ratio the activity decreases at a higher total metal concentration (see entries 1 and 5 in Table 2). ^{13}C NMR data for the polymers obtained are consistent with a branched structure. For this combination, the percentage branching increases with the Fe/Zr ratio. The last columns of Tables 2 and 3 contain the percentage of branching calculated by using a published method.[17]

The polymers produced by $Fe/Me_2Si(Ind)_2ZrCl_2/MAO/C_2H_4$ combinations are characterized by a broad and bimodal molecular weight distribution. The low molecular weight fraction becomes more pronounced as the Fe/Zr ratio increases, giving compositions similar to those of the product obtained with the Fe catalyst. Compared with the product obtained using $Fe/EtInd_2ZrCl_2/MAO/C_2H_4$ systems, the latter are characterized by presenting a monomodal narrow molecular weight. The branching backbone, which increases with increasing concentration of the iron catalyst, and the monomodal weight distribution strongly support the formation of branched polyethylene and an efficient conversion of the monomer to polymeric material.

Heterogeneous Catalysts

In order to turn metallocene catalysts into a morphological control tool, a good method is to heterogenize the catalysts. Although a decrease in catalytic activity with respect to the homogenous systems is expected, the morphology of the polymer and the temperature control of the reaction can be improved. Also, fouling of the reactor, the amount of co-catalyst used, and the cost of the process can be reduced[18-21].

These catalytic systems are based on the immobilization of the metallocene complex on an inert medium (support), without loss of properties and with enhanced physico-chemical characteristics. The basis for these heterogeneous catalysts is to find new supports. These are obtained by preparing different types of oxides like SiO_2, MgO, TiO_2, SiO_2-TiO_2 using sol-gel techniques, leading to more efficient interactions with the metallocene complexes. With this technique it is possible to obtain supports having specific surface area, pore volume and diameter, particle size, and controlled morphology.

Support Preparation

Table 4 summarizes the results of the physical characterization of different silica samples prepared from xerogels via the sol-gel technique. These xerogels were prepared with siloxane sols formed under linear polymerization conditions and by varying the amount of chitosan (CHI) incorporated in them.

Table 4: Effect of precursor xerogel composition, obtained by solvent extraction, on the surface characteristics of silica after calcination for 2 hours at 550 °C.

Sample	Precursor Type	Molar Ratio CHI/SiO$_2$	S$_{BET}$ m^2/g	Pore Diameter Å	Pore Volume cm^3/g	Obs.
S$_0$	Formic Ac	0	45	493	0	Blank
S$_6$	Si sol with chitosan	0.06	323	14	0.156	Microporous
S$_5$	Si sol with chitosan	0.1	327	15	0.16	Microporous
S$_4$	Si sol with chitosan	0.2	503	30	0.221	Microporous
S$_3$	Si sol with chitosan	0.5	632	31	0.41	Mesoporous

S$_{BET}$: Specific surface area.

In sample S$_0$ the silica obtained from a precursor gel without CHI shows a low specific surface area and is non-porous in the measured range. Here, the siloxane sol was mixed with HCOOH which had been used as the solvent for CHI, in order to use the resulting silica as a blank which shows that formic acid does not affect sample porosity. The result is in agreement with the formation of non-porous films in the samples obtained from linear sols.

On the other hand, surface area, as well as pore volume and average pore size, increases with the amount of CHI in the precursor solution up to a CHI/SiO$_2$ molar ratio of 0.5. The pore size distribution is narrow and monomodal, which could be explained on the basis of an homogeneous distribution of the growing polymer domains inside the inorganic network. By studying hybrid CHI/SiO$_2$ films with molar ratios 0.5-0.6 (1:1 by weight), we found that a hybrid complex with a nanocomposite is formed.[22]

The increase of initial amount of chitosan (sample S$_6$), the differential plot for pore volume shows a multimodal and broad distribution of pore size. We think that an excess over the molar composition of the complex should induce phase separation and formation of heterogeneous domains in the composite sample. As expected, after elimination of the organic part the size of the pores should be related to the size of the polymer domains present in the precursor xerogel.

In conclusion, we have shown that the porosity of the silica can be tailored by varying the composition of the hybrid precursor. Moreover, the morphology of the hybrid xerogel and of the final silica after the elimination of the organic residue depends on the method of preparation.[23]

Catalyst Preparation

When the goal is to obtain supports for olefin polymerization, the most important parameters to control are the physicochemical characteristics of the support. Figure 9 shows catalyst preparation with the supports obtained by the sol-gel technique, and Table 5 shows the results of the preparation of the different supports.

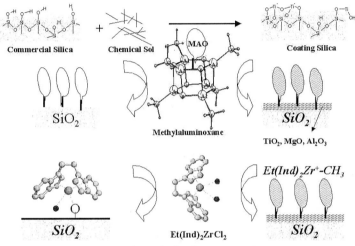

Figure 9: Catalyst preparation.

Table 5 shows the change in the values of the BET area and of the volume and size of the porous supports, by coating commercial silica (ES70) particles with different oxides, and particularly after modification with MAO. This last step, modification with methylaluminoxane, shows the most remarkable changes in the properties of the particles.

The supports presented below had high activities with the Et(Ind)$_2$ZrCl$_2$ (I) catalyst and MAO as a co-catalyst in ethylene polymerization. These catalytic systems reach catalytic activities comparable to that of the commercial silica (SiO$_2$MAO), but require less methylaluminoxane in the modification step (see Table 5). Table 6 summarizes the results of ethylene polymerization with these types of catalysts.

Table 5: Surface modification of silica (ES70) by the sol-gel technique: change in specific surface area, porosity and chemical composition of the supports.

ES70 Coating %	S_{BET} m^2/g	Pore Volume cm^3/g	Pore Diameter Å	Modified Support	S_{BET} m^2/g	Pore Volume m^3/g	Pore Diameter Å	Al %-pp	Ti %-pp
-	255	1,34	171	ES70M	206	1,02	156	10,5	-
Al_2O_3-100	236	1,42	194	ES70AlM	184	0,96	164	-	-
TiO_2-50	226	1,35	187	ES70Ti5M	193	0,87	143	12,8	2,0
TiO_2-100	236	1,38	189	ES70TiM	216	-	-	10,2	4,2
SiO_2-TiO_2-100	249	1,48	191	ES70SiTiM	210	-	-	13,9	1,9
MgO-100	241	1,47	181	ES70MgM	229	0,96	139	-	-
MgO-50	247	1,49	187	ES70Mg5M	-	-	-	-	-
-	-	-	-	SiO_2MAO*	328	0,51	75	22	-
-	-	-	-	TiO_2M **	80	0,15	65	10	-

ES70: Commercial Silica; M: modification with MAO.
 * Witco commercial silica treated with MAO.
** TiO_2, obtained by sol-gel techniques with chitosan.

Table 6: Effect of type of support on catalytic activity in the polymerization of ethylene (Et(Ind)$_2$ZrCl$_2$/MAO).

Catalyst	% Zr	Al/Zr	[Zr] Moles	Activity [Kg PE/ mol h bar]	Bulk Density [g/mL]
SiO$_2$MMEt	0.2	2500	6.E-06	2258	0.22
	0.19	2400	6.E-06	2569	0.24
ESSiTiMMEt	0.18	3000	4.E-06	2919	0.42
	0.22	2500	6.E-06	2342	-
ESTiMMEt	0.2	2400	6.E-06	2150	0.23
	0.2	2400	6.E-06	2036	0.13
ESMgMMEt	0.15	3600	4.E-06	3901	0.40
	0.27	2400	6.E-06	2349	0.18
ESMg5MMEt	0.17	3100	5.E-06	2354	0.22
ESSiTiMMEt 20/80	0.11	3100	5.E-06	3247	-
MgOMMEt	0.2	2500	6.E-06	4032	0.16
S$_0$MMEt*	0.12	2500	6.E-06	3465	0.28
	0.12	2500	6.E-06	3904	0.30

* Correspond to a silica from table 4.

Catalyst SiO$_2$MMEt is well known and is taken as a reference. In general, all the systems studied had similar activities, and many of them had higher activities, as commented above, particularly those catalysts obtained from ESMgMMEt, MgOMMEt and S$_0$MMEt, the latter supported on silica obtained by the sol-gel technique.

The use of these heterogeneous systems has improved polymer morphology because the polymer grows around the support and takes its particle shape. So, we can have a polymer with a spherical shape and a controlled morphology depending on the final use of this material.

In general, the trend in the copolymerization of ethylene with 1-hexene and 1-octadecene in heterogeneous systems is the same for the homogeneous system (see Figure 10). This means that the olefin with the longer chain (1-octadecene) shows the highest activity for both systems, however the catalytic activity values are as expected lower for the heterogeneous systems.

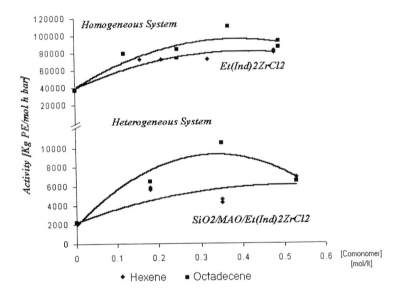

Figure 10: Comparative results between homogeneous and heterogeneous system for the copolymerization of ethylene with 1-hexene and 1-octadecene.

Acknowledgments

The authors acknowledge the financial support of CONICYT through FONDAP Project 1198002, and Witco and Boulder Scientific Company for supplying the MAO and Metallocene, and PETROQUIM for supplying the monomers.

References

[1] VK. Gupta In: Cheremisinoff NP, editor. Handbook of Engineering Polymeric Material, New York: Marcel Dekker, **1997**,155-65

[2] (a) W. Kaminsky, H. Sim, Eds.; Springer-Veralg; Berlin. **1988** (b) G. Fink, R. Mülhaupt, H.H. Brintzinger, Eds.; Springer-Veralg; Berlin. **1995** (c) A. Togm, R.L. Halterman, Eds.; Wiley-VCH: New York **1998**. (d) G.J.P. Britovaek, V.C. Gibson, D.F. Wass, Angew Chem.; Int. Ed. Ed.Engl. **1999**. 429. (e) M.J. Bochmann, Chem. Soc. Dalton trans. **1996**.

[3] (a) D. Rotman, chem. Week **1996**, 158 (36), 37. (b) M. Paige, M. Chem. Eng. News **1998**, 76, 6(49), 25.

[4] H. H. Brintzinger, D. Fischer, R. Mülhaupt, B. Rieger, R. Waymouth, M. Angew. Chem. Int. Ed. Engl. **1995,** 34, 1143a) B. L. Small, M. Brookhart, A. Bennett, J. Am. Chem. Soc. **1998**, 120, 4049. b) R. W. Barnhart, G. C. Bazan, J. Am. Chem Soc. **1998**, 120, 1082. c) W. Kaminsky, R. Engeausen, K. Zoumis, W. Spaleck, J. Rohrmann, Makromol. Chem. **1992**, 193, 1643.

[5] L. Resconi, A. Fait, F. Piemontesi, M. Colonnesi, H. Rychlicki, R. Zeigler, Macromolecules **1995**, 28, 6667-6676.

[6] (a) J. Ewen, J. Am. Chem. Soc. **1984,** 106, 6355. (b) J. Ewen, U.S. Patent **1985**, 4,522, 982 to Exxon. (c) J. Ewen, L. Haspeslagh, J. Atwood, H. Zhang, J. Am. Chem. Soc. **1987**, 109, 6544.

[7] W. Kaminsky, K. Külper, H. H. Brintzinger, F. Wild, Angew. Chem. Int. Ed. Engl. **1985**, 24, 507.

[8] V. Busico, R. Cipullo, Prog Polym. Sci., **2001**, 26, 443.

[9] R. Quijada, G. Galland, RS. Mauler, Macromol Chem Physics **1996**,197, 3091-8.

[10] R. Quijada, A. Narváez, R. Rojas, F.M. Rabagliati, G.B. Galland, R.S. Mauler, R. Benavente, E. Pérez, J. Pereña, A. Bello, Macromol. Chem. Phys. **1999**, 200, 1306.

[11] R. Benavente, E. Pérez, R. Quijada, J Polym Sci B: Polym Phys **2001**,39, 277-285.

[12] J.L. Guevara, R. Quijada, P. Saavedra, H. Palza, G. Galland, submitted to Bol. Soc. Chil. Quím. **2001**.

[13] R. Quijada, J. Dupont, M. Lacerda, R. Scipioni, G. Galland, Macromol. Chem. Phys. **1995,** 196, 3991.

[14] J.L. Guevara, R. Rojas, A. Narvaez, R. Quijada, Bol. Soc. Chil. Quím. **1999**, 44, 497.

[15] J.L. Guevara, M. Yazdani-Pedram, G. Galland, D. Ribeiro, Journal of Polymer Science, Part A: Polymer Chemistry, Submited March, **2002**.

[16] R. Quijada, R. Rojas, G. Bazan, Z. Komon, R. Mauler, G.B. Galland, Macromolecules , **2001**,34, 2411-2417.

[17] G.B. Galland, R. F. de Souza, R.S. Mauler, F. F. Nunes, Macromolecules, **1999**,32, 1620.

[18] T. Kamfjord, T.S. Wester, E. Rytter, Macromol. Rapid Commun. **1998**,19, 505-509.

[19] M.R. Ribeiro, A. Deffieux, M. Portela, Ind. Eng. Chem. Res. **1997**, 36, 1224-1237.

[20] J. Tudor, D. O'Hare, Chem. Commun. **1997**,6, 603.

[21] K. Soga, T. Arai, H. Nozawa, T. Uozomi, Macromol. Symp. **1995**, 97, 53.

[22] J. Retuert, A. Nuñez, M. Yazdani-Pedram, F. Martínez, Macromol. Rapid Commun, **1997**,18,163.

[23] J. Retuert, M. Yazdani-Pedram, V. Arias, R. Quijada, Chemistry Materials, Submitted Nov., **2001**.

The Strange Case of the "Oscillating" Catalysts

Vincenzo Busico*[#], Roberta Cipullo[#], Winfried Kretschmer[§],
Giovanni Talarico[#], Michele Vacatello[#], Valeria Van Axel Castelli[#]

[#] Dipartimento di Chimica, Università di Napoli "Federico II" - Via Cintia - 80126 Naples (Italy)
[§] Department of Chemistry, Rijksuniversiteit Groningen - Nijenborgh 4, 9747 AG Groningen (The Netherlands)

Summary: The field of stereoselective propene polymerization has been dramatically innovated by the discovery of homogeneous metallocene-based catalysts with well-defined and tunable molecular structure. Of all, "oscillating" metallocenes are probably the most ingenious and challenging example of catalyst design. Their catalytic species were built to "flip-flop" between a chiral and an achiral conformation, at a rate intermediate between those of monomer insertion and chain transfer. The result of this molecular switching would be a polypropylene with an isotactic/atactic stereoblock structure, performing as a thermoplastic elastomer. This essay discusses how the real polymerization mechanism differs from what the catalyst inventors had in mind, but also how - through fortunate circumstances - their optimism has been rewarded.

Introduction

The advent, in the mid-'80s, of metallocene-based catalysis was a true revolution in the world of stereoselective olefin polymerization.[1,2] After thirty years since the initial discoveries of Karl Ziegler and Giulio Natta,[3,4] and five generations of increasingly effective but poorly understood heterogeneous catalysts,[5] it was ultimately found that homogeneous catalytic species with well-defined precursors and a predetermined ligand environment of the active transition metal can promote with high productivity the regio- and enantioselective polyinsertion of 1-alkenes.

Polypropylene synthesis, in particular, became a favorite playground (which is understandable, in view of the enormous industrial significance of this resin[5]). In a comparatively short time, classes of metallocene catalysts were developed with almost any conceivable kind of stereocontrol (isotactic, syndiotactic, atactic, hemiisotactic), and - most importantly - the relationship between selectivity on one

128

hand, and symmetry and structure of the catalytic species on the other, was clarified.[1,2]

All of a sudden, the dream to tailor a catalytic species to a desired polymer architecture looked close to reality. Therefore, when a rational route to isotactic/atactic stereoblock polypropylene (a material of high potential interest for applications as a thermoplastic elastomer) was announced,[6] the scientific community was - in a way - prepared and there was little room for skepticism.

The idea behind the catalyst design was elegant and conceptually simple (Chart 1 - P = Polymer chain; only C-C bonds traced in the saw-horse chain representations). As is well-known, stereorigid *ansa*-metallocene catalysts in which two indenyl ligands are locked by a bridge in a *rac*-C_2-symmetric (Chart 1-a) or a *meso*-C_s symmetric (Chart 1-b) configuration afford isotactic and atactic polypropylene, respectively.[1,2] The former is a semicrystalline thermoplastic material, with a melting temperature of up to 165°C; the latter, instead, is uncrystallizable and moderately elastomeric.[5] With this premise, how about preparing *unbridged* bis-indenyl catalysts with substituents of tunable size on the rings, in such a way that hindered ligand rotation is allowed, and a *rac/meso* conformational rearrangement occurs at a rate intermediate between those of monomer insertion and chain growth (transfer)?

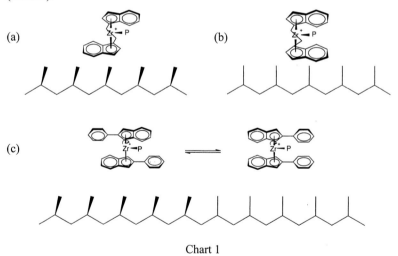

Chart 1

The concept substantiated in the form of (2-Ar-Indenyl)$_2$ZrCl$_2$ complexes, where Ar is an aryl group which can range from a simple phenyl (**1**; Chart I-c)[6] to much more complicated and bulky moieties, like 3,5-di-*tert*-butyl-4-methoxy-phenyl (**2**; Chart 2).[7] Single crystal X-ray diffraction had proved that complex **1** crystallizes in mixed "*rac*-like" and "*meso*-like" conformation.[6] When combined in toluene solution with methyl-aluminoxane (MAO) or another suitable activator,[1,2] it gives rise to a propene polymerization catalyst of decent activity. Remarkably, the polypropylene produced is largely stereoirregular, but also contains a highly isotactic part;[6,7] the fact that it does perform as a thermoplastic elastomer[6-8] was taken as an indication that at least part of the isotactic and atactic sequences are chemically bound, and that crystalline domains act as physical crosslinks between amorphous ones.[8] In turn, this was perceived as a validation of the mechanistic picture of Chart 1-c, and the whole class of (2-Ar-Indenyl)$_2$ZrCl$_2$ catalysts gained the now popular definition of "oscillating", and were credited as clean "molecular switches".[6-10]

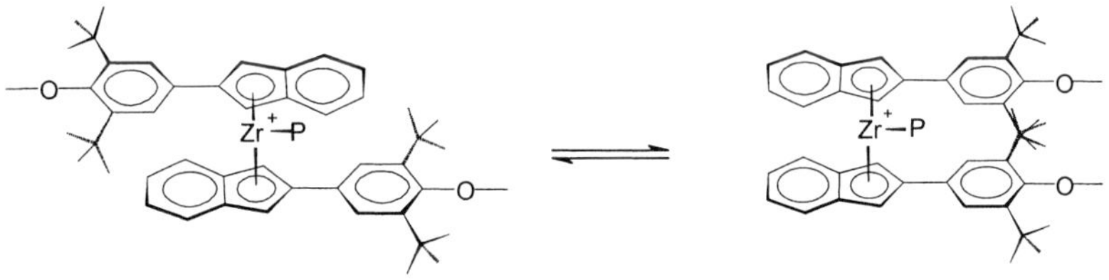

(a)   (b)

Chart 2

However, the following facts should have suggested that the reality is somewhat more complicated:

i) Solution NMR investigations on model complexes, down to very low temperatures, did not provide evidence for a slow *rac*/*meso* interconversion .[11]

ii) Although even small changes in the ligand framework can affect the stereoselectivity dramatically, no clear structure/performance correlation could be found.[7]

iii) Computer modeling by means of molecular mechanics, alone[12] or combined with quantum mechanics,[13] confirmed that, for complex 1, the *rac*-like and *meso*-like conformations correspond to energy minima (the former being slightly deeper), but the activation energy for their interconversion was estimated to be fairly low (2-5 kcal/mol),[13] and - at most - comparable with that for monomer insertion (5-15 kcal/mol).[1] Considering that the latter is a bimolecular process (hence with a large negative activation entropy), it is difficult to imagine how it can be (much) faster than the intramolecular *rac*-like/*meso*-like rearrangement. Higher interconversion barriers were calculated for complexes with bulkier Ar substituents, such as 2,[14] but in that case the *meso*-like conformation turned out to be much less stable than the *rac*-like one, due to unfavorable contacts between the Ar groups (apparent already on inspection of Chart 2-b).

iv) Last but not least, the polymers have broad molecular mass distributions, which is rather unusual with homogeneous catalysts; moreover, they can be solvent separated in fractions largely differing in stereoregularity, from completely amorphous (though not purely atactic) to highly crystalline (though not completely isotactic).[2,7,8] Altogether, this strongly suggests the presence of more than one catalytic species, possibly in equilibrium but with average life-times longer than the average growth time of individual macromolecules.

On the other hand, direct proof (or disproof) of the mechanism in Chart 1-c by means of a microstructural characterization of the polymers was never provided. That is what we decided to achieve.

Reading "oscillating polypropylenese"

Stereoselective polymerization is a very special case of asymmetric synthesis. For a reaction affording low-molecular-mass products, the effectiveness of the asymmetric induction can be measured in terms of the so-called enantiomeric excess (*i.e.*, the difference in fractional abundance of the two resulting enantiomers), but this is practically uninformative on the reaction mechanism. For a polymerization, instead, the fact that the products of individual reaction steps (*i.e.*, the monomeric units) are permanently enchained in the form of macromolecules represents an extraordinary advantage for mechanistic purposes. In fact, from the stereochemical characterization of a polymer, in addition to the degree of stereoregularity (which is the equivalent of the enantiomeric excess), the way each reaction step affected the subsequent one(s) can also be derived. In this respect, a polymer chain contains its complete birth story, sequentially recorded like on a tape; of course, to know that story one must be able to read the tape.

For decades, we Ziegler-Natta chemists have tried to do so with polypropylene, in order to trace the behavior of our complicated and molecularly ill-defined heterogeneous catalysts. Year after year, our reading has become more and more fluent, and our understanding lately deeper and deeper.[15,16]

"Polypropylenese" is a rather weird language, based on two letters only - namely, "*m*" and "*r*" (Chart 3-a) - but with a fairly high number of local dialects. The isotactic version, in particular, makes use predominantly of *m*'s in long strings, with just a few *r*'s usually occurring in couples (*rr*) (Chart 3-b). The atactic form, instead, is much more varied, with an equal number of randomly distributed *m*'s and *r*'s (Chart 3-c).

Understanding the two is not difficult, unless they are intermixed, which is unfortunately the case for polypropylene made with oscillating metallocenes. In particular, to solve the problem introduced in the previous Section one has to judge whether whole polymer chains, or - rather - different segments within individual chains, are written in the two idioms. This requires much more than merely counting the characters (*m* and *r* diads), or even the syllables (*mm*, *mr* and *rr* triads); in fact, one has to look at whole words, and to check for the possible presence of hybrid ones (Chart 3-d).

Our reading eye for "polypropylenese" is ^{13}C NMR.[15] Routine ^{13}C NMR spectra of polypropylene give easy access to the distribution of sequences with up to four characters (pentads; Figure 1, top), but this is not enough yet for the considered case.[9,17] In our laboratory, however, we have recently developed high field (150 MHz) ^{13}C NMR techniques enabling us to go further - in favorable cases up to twelve characters, more typically from six to eight (Figure 1, bottom).[15,16,18]

132

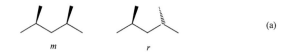

(a)

m   *r*

(b)

m   *m*   *m*   *r*   *r*   *m*   *m*

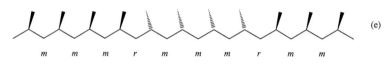

(c)

m/r   *m/r*   *m/r*   *m/r*   *m/r*   *m/r*   *m/r*

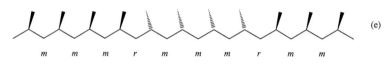

(d)

m   *m*   *m*   *m*   *m/r*   *m/r*   *m/r*   *m/r*

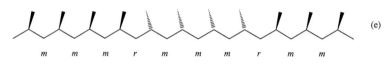

(e)

m   *m*   *m*   *r*   *m*   *m*   *m*   *r*   *m*   *m*

Chart 3

When we applied this tool to samples of polypropylene made with two representative oscillating catalysts, namely **1** and **2** (activated at 20°C in toluene solution with typical co-catalysts such as MAO or N,N-dimethylanilinium *tetrakis*-perfluorophenylborate/Al(*iso*-Butyl)₃)[1,2] what we read was partly unexpected.

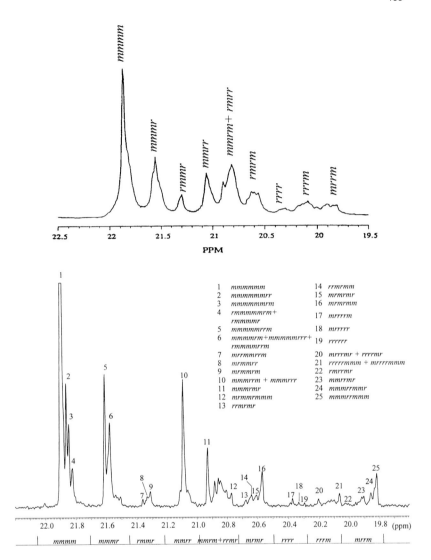

Figure 1. Methyl region of the ^{13}C NMR spectra of typical polypropylene samples prepared with catalyst **1**. Top - 100 MHz (adapted from ref. *9*). Bottom - 150 MHz (this work). The chemical shift scale is in ppm downfield of tetramethylsilane. Thanks to a very high resolution and a full resonance assignment, in the 150 MHz spectrum it was possible to identify for the first time peaks (no. 11 and 16) mostly due to junctions between isotactic and (quasi-)atactic blocks.[19]

In agreement with the literature,[2,7,8] the polymers obtained with **1** could be solvent-separated in fractions according to crystallinity, from nil up to 60%, as measured by X-ray diffraction (Table 1, entries 1-4). In all fractions we found isotactic segments (with traces of *rr* stereodefects), along with largely stereoirregular ones, and it was gratifying for us to be the first to recognize, in the semicrystalline fractions, points of switch between the two (*e.g.*, the intense *mmmrmr* and *mrmrmm* methyl heptads; Table 2 and Figure 1-right).[19] To our surprise, however, we also noted that the non-crystallizable blocks were not truly atactic, but contained a slight excess of *m*'s, although still distributed at random as the atactic grammar requires;[15] this feature was more pronounced the higher was monomer concentration in the feed (compare the values of the P_m parameter for entries 1,2 and 3,4 in Table 1).

Table 1. Results of propene polymerization in the presence of the "oscillating" metallocene catalysts **1** and **2** (see text)

Entry no.	Catalyst/ Co-catalyst	Solvent	$[C_3H_6]$, mol/L	[*mmmmmm*], % [a]	P_m [b]	T_m, °C [c]	Δh_m, J/g [c]
1	**1/MAO**	Toluene	1.5	5	0.56	-	-
2	**1/MAO**	Toluene	6.7	8	0.58	-	-
3	**1/Borate** [d]	Toluene	1.5	13	0.53	133	7
4	**1/Borate** [d]	Toluene	6.7	23	0.59	140	10
5	**2/MAO**	Toluene	1.5	30	0.84	90	4
6	**2/MAO**	Toluene	6.7	52	0.91	140	34
7	**1/Borate** [d]	ODCB [e]	1.5	4	0.52	-	-

Experimental conditions: $T = 20°C$; $[Zr] = 3×10^{-5}$ mol/L; $[Al]/[Zr] = 1.0×10^3$ (for activation with MAO); $[B]/[Zr] = 2.5$, $[Al]/[Zr] = 3.0×10^2$ (for activation with dimethylanilinium *tetrakis*-perfluorophenylborate/Al(*iso*-Butyl)₃); polymerization time, 1.00 h.

[a] ^{13}C NMR fraction of isotactic heptads in the whole sample.
[b] Conditional probability of *m* diad formation during stereoirregular chain propagation, estimated by statistical analysis of the ^{13}C NMR stereosequence distribution.[15]
[c] Measured by Differential Scanning Calorimetry (DSC) on 2nd heating scan.
[d] *tetrakis*-perfluorophenylborate/Al(*iso*-Butyl)₃).
[e] 1,2-dichlorobenzene.

Table 2. 150 MHz ^{13}C NMR stereosequence distribution of the diethyl-ether-soluble/hexane- insoluble fraction of a polypropylene sample prepared with catalyst **1** (Table 1, entry 4), and best-fit calculated one in terms of two stochastic models: one assuming the sample as a physical mixture of isotactic and (quasi-)atactic chains, another as an isotactic/(quasi-)atactic stereoblock polymer.[9,15,16,19] The latter model ends up with a much better agreement, as the reduced-χ^2 test indicates.

Stereosequence	Normalized fraction		
	Experimental	Calculated (physical mixture)	Calculated (stereoblock)
mmmm	0.4854(63)	0.4833	0.4875
mmmmmm	0.3808(66)	0.3832	0.3752
mmmr	0.1453(20)	0.1437	0.1446
mmmmrr	0.0716(39)	0.0705	0.0724
rmmr	0.0212(20)	0.0236	0.0197
mmrr	0.0990(20)	0.1022	0.0989
mmrm+rmrr	0.1096(20)	0.1149	0.1098
mmmrmr	0.0330(20)	0.0226	0.0324
rmrm	0.0521(20)	0.0472	0.0504
mrmrmm	0.0288(20)	0.0226	0.0306
rrrr	0.0093(20)	0.0078	0.0089
rrrrrm	0.0030(20)	0.0027	0.0030
rrrrrr	0.0012(10)	0.0008	0.0009
rrrm	0.0278(20)	0.0262	0.0285
rmrrmr	0.0030(20)	0.0025	0.0019
mmrrmr	0.0128(20)	0.0133	0.0121
mmrrmm	0.0345(20)	0.0353	0.0336
		Isotactic content = = 57 wt.-% $\chi_r^2 = 4.0$	$P_{iso/ata} = 0.19$ [a] $P_{ata/iso} = 0.33$ [b] $\chi_r^2 = 0.3$

[a] Conditional probability of switching from isotactic to (quasi-)atactic propagation.
[b] Conditional probability of switching from (quasi-)atactic to isotactic propagation.

Even more surprisingly, in polypropylene samples prepared with catalyst **2** the chains turned out to be largely written in a more exotic dialect, made of strings of consecutive *m*'s mostly separated by randomly distributed *single r*'s (Chart 3-e and

Table 3); this can be seen on inspection of the high-field ^{13}C NMR spectrum of Figure 2, showing an intense sharp peak corresponding to the *mmmrmm* heptad.

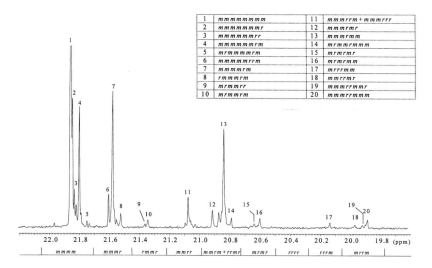

1	*mmmmmmm*	11	*mmmrrm + mmmrrr*
2	*mmmmmmr*	12	*mmmrmr*
3	*mmmmmrr*	13	*mmmrmm*
4	*mmmmmrm*	14	*mrmmrmm*
5	*mrmmmmrm*	15	*mrmrmr*
6	*mmmmrm*	16	*mrmrmm*
7	*mmmmrm*	17	*mrrrmm*
8	*rmmmrm*	18	*mmrrmr*
9	*mrmmrr*	19	*mmmrrmmr*
10	*mrmmrm*	20	*mmmrrmmm*

Figure 2. Methyl region of the 150 MHz ^{13}C NMR spectrum (in 1,2-tetrachloroethane-d_2 at 90°C) of a polypropylene sample prepared with catalyst **2** (Table 1, entry 6). The chemical shift scale is in ppm downfield of tetramethylsilane; the assignment is based on Ref. 13. The presence in the sample of isotactic sequences separated by *single r* diads is apparent already on inspection (see, in particular, the intense peaks numbered as 4, 7 and 13).

We can view this as an extreme degeneration of the atactic form, exaggerating the deviation just described for catalyst **1** and thus ending up with a much larger defect of *r*'s. As a matter of fact, in this case too the fraction of *m*'s turned out to increase with increasing propene concentration in the feed (Table 1, entries 5 and 6).

All this combines in an intriguing story, which proves - if needed - that science and common sense are not always good friends.

Table 3. 150 MHz ^{13}C NMR stereosequence distribution of a polypropylene sample prepared with catalyst **2** (Table 1, entry 6), and best-fit calculated one in terms of a stochastic model assuming chain propagation at a *rac*-like catalytic species oscillating between its two enantiomorphous forms (see text).

Stereosequence	Normalized fraction	
	Experimental	Calculated
mmmm	0.6446(125)	0.6495
mmmmmm	0.5178(127)	0.5266
mmmmmr	0.1082(30)	0.1165
mmmr	0.1419(22)	0.1436
mmmmrr	0.0255(20)	0.0267
rmmr	0.0105(20)	0.0079
mmrr	0.0342(20)	0.0329
mmrm+rmrr	0.1336(22)	0.1303
mmmrmm	0.0991(20)	0.1025
mmmrmr	0.0147(22)	0.0116
rmrm	0.0162(20)	0.0143
mrmrmm	0.0098(20)	0.0102
rrrr	0.0009(20)	0.0005
rrrm	0.0045(20)	0.0053
mrrm	0.0145(20)	0.0156
mmrrmr	0.0029(20)	0.0028
mmrrmm	0.0116(20)	0.0127
		$P_{osc} = 0.086$ [a]
		$\sigma = 0.985$ [b]
		$\chi_r^2 = 1.3$

[a] Conditional probability of oscillation of the catalytic species.
[b] Probability to select the preferred monomer enantioface at each enantiomorphous catalytic species.

From microstructure to mechanism

The presence of isotactic sequences with minor amounts of *rr* stereodefects (Chart 3-b) in polypropylene samples made with (2-Ar-Indenyl)$_2$ZrCl$_2$ catalysts proves unquestionably that the catalytic species can exist in solution in the *rac*-like conformation for a time longer than that required for a single monomer insertion.

Conversely, the genesis of stereosequences with an excess of *m* diads separated by *single r* diads (Chart 3-e) is more ambiguous. In principle, they might be ascribed to chain propagation under the chiral control of the growing chain end (1,3-*like* asymmetric induction)[15,20] when the catalytic species is in the achiral *meso*-like conformation. In such a case an occasional failure of stereocontrol, inverting the configuration of the last inserted unit, would be perpetuated until another failure takes place. Sporadic examples of weak isotactic chain-end control have actually been reported for metallocene catalysts with non-chirotopic sites,[1,2,15,21] but only with bis-cyclopentadienyl ligands and at very low temperature (<<0°C); moreover, a dependence of the extent of chain-end control on monomer concentration has never been observed, at least to our knowledge.

A much more plausible hypothesis is that such blocks are formed instead at a *rac*-like catalytic species which oscillates between its two enantiomorphous forms (rather than to and from the less stable *meso*-like form), at a rate (slightly) lower than that of monomer insertion (Chart 4-bottom).[19] Indeed, as is well known for the *rac-ansa*-metallocene homologues (Chart 1-a), opposite monomer enantiofaces are inserted at active species with opposite chirality.[1,2] This offers an immediate and simple explanation for the otherwise puzzling increase of the *m* fraction (P_m) - or, equivalently, of the average length of the m_n strands - with increasing monomer concentration (because that speeds up the polyinsertion, but not the oscillation).

From Table 1, one has to conclude that the ratio between the average rates of monomer insertion and ligand rotation (which is roughly coincident with $P_m/(1-P_m)$) is critically dependent on the Ar substituent. In particular, it can be comparatively high ($P_m > 0.90$) for the bulky catalyst **2**, whereas for the much less hindered catalyst **1**, in the time scale of monomer insertion, ligand rotation can be regarded as almost free even in liquid propene, and the result is a quasi-atactic chain propagation ($0.50 < P_m < 0.60$. This obviously implies that, for the latter system, some contribution of the *meso*-like form to the catalytic activity cannot be ruled out). All this fits with the previously discussed results of computer modeling.[12-14]

Now we need to explain why a highly fluxional catalytic species can be frozen occasionally in one of the two mirror images of its *rac*-like conformation for a comparatively long time, and thus produce (long) isotactic blocks. An important

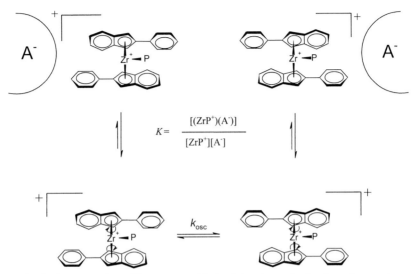

$$K = \frac{[(ZrP^+)(A^-)]}{[ZrP^+][A^-]}$$

clue is the co-catalyst effect: we noted indeed that, in particular for polypropylene

Chart 4

made with catalyst **1**, changing MAO for N,N-dimethylanilinium *tetrakis*-perfluorophenylborate/Al(*iso*-Butyl)₃ results (*coeteris paribus*) in a significantly higher content of isotactic blocks (Table 1, entries 1,3 and 2,4). Keeping in mind that: i) the active species of metallocene catalysts are cationic;[1,2] ii) the co-catalyst has to be a "poorly coordinating" anion, in order not to form a tight ion couple that would preclude the access to the monomer;[1] and, iii) the polymerization is normally carried out in non-polar or weakly polar solvents (such as toluene or hexane), in which catalyst and co-catalyst are known to give rise to a variety of "loose" ion couples (monomer-separated, solvent-separated, etc.),[1,2,22] we conclude that it is the interaction with the counterion to freeze, in some of the possible arrangements, the oscillation of a *rac*-like [(2-Ar-Indenyl)₂ZrP]⁺ cation (Chart 4-top). In this respect *tetrakis*-perfluorophenylborate, which can establish directional interactions with metallocene cations, seems to be more effective than the large and highly delocalized anion of MAO.[22,23]

If such hypothesis is correct, one has to anticipate that an increase in the polarity of the reaction medium will result in a decrease of what, in an admittedly crude attempt of simplification, can be defined as "overall association constant" K (Chart 4), and therefore of the isotactic content in the polymerization product. **This is actually the case.** In fact, we carried out two propene polymerizations with catalyst system **1**/N,N-dimethylanilinium *tetrakis*-perfluorophenylborate/Al(*iso*-Butyl)$_3$, under identical conditions except that toluene (dielectric constant, $\varepsilon = 2.38$ at 25°C) in one case, and 1,2-dichlorobenzene ($\varepsilon = 9.93$) in the other was used as the solvent (Table 1, entries 3 and 7); in the former case, the polymer turned out to be semicrystalline, with a fraction of isotactic heptads [*mmmmmm*] = 13% and a melting temperature $T_m \approx 133°C$, whereas in the latter **it was completely amorphous and atactic-like** ([*mmmmmm*] = 4%).

Moral

In conclusion, we believe that the true picture of an oscillating metallocene catalyst is the one shown in Chart 4. If one compares it with that in Chart 1,[6] one can realize how different (and also, unfortunately, more complicated) it is in molecular terms: rather than involving a *rac*-like and a *meso*-like conformation, the "oscillation" takes place (mainly) between the two enantiomorphous forms of the *rac*-like conformation, provided that the counterion leaves room for that. It is unquestionable that the Ar substituent on the indenyl ligands crucially drives catalyst dynamics; however (apart from possible electronic effects) changing its bulkiness affects, at the same time, the relative stability of the accessible conformations, their interconversion barriers and the cation/anion interlocking, which explains why it is hard not only to predict, but even to rationalize *a posteriori* the result on catalyst selectivity.

Therefore, if this story has a moral, we would say that it is close to the old saying that "Fortune favors the brave", and that the way to rational catalyst design can still be long, even for the "simple" metallocenes.

Acknowledgements. This study was funded by the Dutch Polymer Institute (DPI - Project #100). V.V.A.C. acknowledges the DPI for a post-doctoral fellowship. V.B. and R.C. are grateful to the Italian Ministry for University (PRIN 2000) for financial assistance. The authors thank Prof. A.L. Segre and Dr. P. Budzelaar for precious discussions, and Ms. Valentina Langella for sample preparation and fractionation.

[1] H. H. Brintzinger, D. Fischer, R. Mülhaupt, B. Rieger, R. M. Waymouth, *Angew. Chem. Int. Ed. Engl.* **1995**, *34*, 1143.

[2] L. Resconi, L. Cavallo, A. Fait, F. Piemontesi, *Chem. Rev.* **2000**, *100*, 1253.

[3] J. Boor, Jr., *Ziegler-Natta Catalysts and Polymerizations*, Academic Press, New York, **1979**.

[4] Y. V. Kissin, *Isospecific Polymerization of Olefins*, Springer-Verlag, New York, **1985**.

[5] E. P. J. Moore, *Polypropylene handbook: polymerization, characterization, properties, applications*, Hanser Publishers, Munich, **1996**.

[6] G. W. Coates, R. M. Waymouth, *Science* **1995**, *267*, 217.

[7] a) E. Hauptman, R. M. Waymouth, J. W. Ziller, *J. Am. Chem. Soc.* **1995**, *117*, 11586; b) R. Kravchenko, A. Masood, R. M. Waymouth, *Organometallics* **1997**, *16*, 3635; c) J. L. Maciejewski Petoff, M. D. Bruce, R. M. Waymouth, A. Masood, T. K. Lal, R. W. Quan, S. J. Behrend, *Organometallics* **1997**, *16*, 5909; d) S. Lin, E. Hauptman, T. K. Lal, R. M. Waymouth, R. W. Quan, A. B. Ernst, *J. Mol. Catal. A* **1998**, *136*, 23; e) P. Witte, T. K. Lal, R. M. Waymouth, *Organometallics* **1999**, *18*, 4147.

[8] a) Y. Hu, M. T. Krejchi, C. D. Shah, C. L. Myers, R. M. Waymouth, *Macromolecules* **1998**, *31*, 6908; b) E. D. Carlson, M. T. Krejchi, C. D. Shah, T. Terakawa, R. M. Waymouth, G. G. Fuller, *Macromolecules* **1998**, *31*, 5343.

[9] M. D. Bruce, R. M. Waymouth, *Macromolecules* **1998**, *31*, 2707.

[10] a) M. Nele, S. Collins, J. C. Pinto, M. L. Dias, S. Lin, R. M. Waymouth, *Macromolecules* **2000**, *33*, 7249; b) S. Lin, C. D. Tagge, R. M. Waymouth, M. Nele, S. Collins, J. C. Pinto, *J. Am. Chem. Soc.* **2000**, *122*, 11275.

[11] a) S. Knüppel, J. Fauré, G. Erker, G. Kehr, M. Nissinen, R. Fröhlich, *Organometallics* **2000**, *19*, 1262; b) T. Dreier, G. Erker, R. Fröhlich, B. Wibbeling, *Organometallics* **2000**, *19*, 4095.

[12] L. Cavallo, G. Guerra, P. Corradini, *Gazz. Chim. It.* **1996**, *126*, 463.

[13] M. A. Pietsch, A. K. Rappé, *J. Am. Chem. Soc.* **1996**, *118*, 10908.

[14] A. Maiti, M. Sierka, J. Andzelm, J. Golab, J. Sauer, *J. Phys. Chem. A* **2000**, *104*, 10932.

[15] V. Busico, R. Cipullo, *Prog. Polym. Sci.* **2001**, *26*, 443.

[16] V. Busico, R. Cipullo, G. Monaco, G. Talarico, M. Vacatello, J. C. Chadwick, A. L. Segre, O. Sudmejier, *Macromolecules* **1999**, *32*, 4173.

[17] W. J. Gauthier, S. Collins, *Macromolecules* **1995**, *28*, 3779.

[18] a) V. Busico, R. Cipullo, P. Corradini, L. Landriani, M. Vacatello, A. L. Segre, *Macromolecules* **1995**, *28*, 1887; b) V. Busico, R. Cipullo, G. Monaco, M. Vacatello, A. L. Segre, *Macromolecules* **1997**, *30*, 6251; c) V. Busico, R. Cipullo, G. Monaco, M. Vacatello, J. Bella, A. L. Segre, *Macromolecules* **1998**, *31*, 8713.

[19] V. Busico, R. Cipullo, A. L. Segre, G. Talarico, M. Vacatello, V. Van Axel Castelli *Macromolecules* **2001**, *34*, 8412.

[20] F. A. Bovey, G. V. D. Tiers, *J. Polym. Sci.* **1960**, *44*, 173.

[21] J. A. Ewen, *J. Am. Chem. Soc.* **1984**, *106*, 6355.

[22] E. Y.-X. Chen, T. J. Marks, *Chem. Rev.* **2000**, *100*, 1391.

[23] a) E. Zurek, T. K. Woo, T. K. Firman, T. Ziegler, *Inorg. Chem.* **2001**, *40*, 361; b) M. Ystenes, J. L. Eilertsen, J. Liu, M. Ott, E. Rytter, J. A. Støvneng, *J. Polym. Sci. Part A* **2000**, *38*, 3106; c) R. Fusco, L. Longo, F. Masi, F. Garbassi, *Macromolecules* **1997**, *30*, 7673.